Phyllotaxis Models

Series Editor
Marie-Christine Maurel

Phyllotaxis Models

A Tool for Evolutionary Biologists

Jean-Paul Walch
Solange Blaise

iSTE

WILEY

First published 2023 in Great Britain and the United States by ISTE Ltd and John Wiley & Sons, Inc.

ISTE Ltd
27-37 St George's Road
London SW19 4EU
UK

www.iste.co.uk

John Wiley & Sons, Inc.
111 River Street
Hoboken, NJ 07030
USA

www.wiley.com

Library of Congress Control Number: 2023933897

British Library Cataloguing-in-Publication Data
A CIP record for this book is available from the British Library
ISBN 978-1-78630-913-6

Contents

Foreword

I could not think of any better circumscription of the concept of phyllotaxis than the one provided by Adler et al. (1997) in their article entitled "A history of the study of phyllotaxis":

> Phyllotaxis, a subdivision of plant morphology, is the study of the arrangement of repeated units such as leaves around a stem, scales on a pine cone or on a pineapple, florets in the head of a daisy, and seeds in a sunflower. Remarkably, these units often form systems of spirals or helices. As a study of the self-organization of repeated units, it somewhat resembles the study of crystallography. In fact, the objects studied may be thought of as living crystals. However, the fact that they are living and growing adds a further dimension to the subject, making it necessary to take into account such things as growth rates, heredity, interaction with the environment, and evolution. Trying to explain the regularity of the arrangements of petals, leaves, etc. has led research workers to incorporate molecular biology, general comparative morphology, and evolutionary theory into their thinking.

Studying phyllotaxis – its causes, its establishment during development and its functions – is a challenge that several scientists and brilliant minds have faced in the past. Historically, Theophrastus,

Pliny the Elder, Leonardo Da Vinci, Johannes Kepler and Johann Wolfgang von Goethe are among the most influential figures who played a pioneering role in the study of phyllotaxis, paving the way to modern studies. Numerous hypotheses have been tested, and many articles have been published in the 19th and 20th centuries, trying to bring forward our knowledge about the mechanisms underlying the different phyllotactic types. This book from Walch and Blaise is a summary of the research conducted so far on the topic, and it also introduces an elegant model for reproducing (accurately) and predicting (quite satisfactorily) phyllotactic patterns.

I assume there will be at least three ways of reading this book, depending on how we can handle the profuse information presented in the seven chapters. While some will be curious about the mathematical and theoretical modeling aspects that are indicative of crucial properties of plant structure and development, others will be interested in the history of paradigm shifts that have characterized the evolution of plant morphology. A third kind of reader will use a taxonomical map to circulate through the landscape of this book and use particular species or supraspecific lineages as magnifying lenses.

Jean-Paul Walch is not a trained botanist. Far from being a drawback, this is what probably allowed him to assertively overcome the supposedly daunting task of modeling phyllotaxis in (mostly) angiosperm lineages, focusing on (mostly) reproductive units, to explain adult (with a static model) and developing (with a dynamical model) structures, relying on concepts from the fields of mathematics, physics, mechanics, evo-devo and molecular biology.

Although the initial project of Jean-Paul Walch and the botanist Solange Blaise was indisputably ambitious, they succeeded in formalizing their ideas and opinions in a book that will for sure inspire researchers from the above-mentioned disciplines and give grist to the mill of anyone who wants to discuss their model and propose adjustments, albeit improvements.

This book will find its place in the library of botanists interested in the theoretical aspects of plant shape and organization, and more broadly of scientists who are keen on deciphering and understanding natural forms and the phenotypes realized in the course of evolution, compared to those that are theoretically possible.

Florian JABBOUR
Associate Professor,
Muséum National d'Histoire Naturelle,
Paris, France

Preface

> "What theory is there in botany that is vast enough to embrace all the facts, and a formula that is flexible enough to counter all objections?"
>
> "Some may thank us for attempting something so foolhardy, without concealing from us the pitfalls of this perilous enterprise."
>
> "The plant kingdom loses itself like an atom in the vast plane of possible systems, but it displays nonetheless a unicity of form in its origin and infinite variation in its details."

– Bravais and Bravais (1830)
"Essai sur la disposition générale des feuilles rectirisées"
(Essay on the general arrangement of whorled leaves).

Phyllotaxis, as a discipline, uses a mathematical approach to understand plant structure. Its conceptual level was already advanced in the early 19th century, notably with the contribution of Auguste Bravais. The time has now come for this concept to become a real tool for biologists. The few equations that describe in a general, and thus simplified manner the biochemical and physical mechanisms underlying the development of each species can only be solved with numerical calculations, that is, using computational means.

This approach may be offputting for biologists. We therefore suggest they draw information from this book in the following manner:

– Chapter 1 presents the main concepts of phyllotaxis: parastichies, divergence angles, plastochron ratios, etc., which are systematically illustrated with examples from the botanical world.

– Chapter 2 retraces the history of this discipline without going into any mathematical detail, allowing the reader to understand the evolution of the theoretical aspects of phyllotaxis, and highlights the turning point that was Douady and Couder's "standard model" in 1996.

– Chapter 3 presents the model of the classic sunflower capitulum (the "static model"), showing that Fibonacci spirals optimize organ packing even though they are not the only configuration to do so. Although this idea is met with some resistance by those who see it as a finalist approach, Fibonacci spirals minimize the self-shading of leaves. Biologists may limit themselves to the discussion, and perhaps read the section on the golden ratio, which was based on a high school project.

– Chapters 4 and 5 are essentially mathematical: they present the equations of Douady and Couder describing the arrangement of organs in developing meristems, as well as the parametrization we used and with which we were able to model the diversity of the arrangements found in plants. Indeed, our predecessors who developed this model relied on too few parameters, like meteorologists who only have data from a few recording stations over an area as large as France. Biologists can nevertheless read our presentation of the model of molecular phenomena underlying plant phyllotaxis. On request, the authors can provide a simple Excel spreadsheet where the initial position of two to four primordia can be set, and the positions of the following six primordia can be calculated[1].

1 To request the Eichler.xls spreadsheet please contact Jean-Paul Walch at jpwalch511@gmail.com.

– Chapter 6 presents the evolution of Magnoliales and Laurales in a new light. Botanists can simply read the discussion and refer to the main text for details of how the results were obtained. If the model described at the start of the chapter allows a new interpretation of particular arrangements in members of these orders, its most significant result is the concept of pseudo-whorls. Although flowers in basal angiosperms are either completely spiraled or whorled, flowers in Magnoliales and especially Laurales evolved the spiral organization to acquire qualities that were unique to the whorls. This explains why the numbers of organs in these pseudo-whorls, i.e. 3, 5, etc. belong to the Fibonacci sequence!

– Chapter 7 presents the organization of inflorescences. It does not rely on mathematics, but is based on the elementary geometrical construction of inflorescences, namely, the assembly of “bricks” consisting of floral meristems and their bracts and bracteoles.

The conclusion is purely textual.

Our model has been published in the *Journal of Theoretical Biology*, its application in basal angiosperms in *Botany Letters* and the demonstration of the influence of bracteoles on sepal arrangement (initially intuited by Eichler in 1875) in *Flora*.

February 2023

Acknowledgments

We would like to thank Florian Jabbour, Associate professor, Muséum National d'Histoire Naturelle, Paris, for his encouragement, Hélène Citerne for her review of the manuscript and Line Walch for her superb drawings.

Introduction

After trying for over a century and a half, scientists have discovered the equations of the phyllotaxic engine: these describe the inhibition "force" generated by incipient primordium preventing the initiation of other primordia in its immediate vicinity (Douady and Couder 1996a). What was still lacking was a method to specify this notion of proximity.

The very precise measurements of the vegetative meristem of *Linum usitatissimum* made by Williams (1975) showed us how to proceed: the method that was missing was to take into account the distance of primordia to the meristem apex during its development.

We consider that our mathematical model with these parameters is an instrument for evaluating the "forces" in (biochemical and mechanical) and on (pressure) the meristem from their effect on the geometry of this meristem. With this powerful tool, we have revisited many important notions of plant biology.

Spiral phyllotaxis, be it Fibonacci, Lucas or of another type, optimizes organ compaction (see Chapter 3). By contrast, whorled phyllotaxis is suboptimal for this feature, but is associated with clear organ differentiation.

Our approach allowed us to precisely model the development of the vegetative meristem of *Linum usitatissimum* from the two cotyledons up to the initiation of 80 leaves arranged in a Fibonacci spiral (see Chapter 4).

This model can also be applied to the initiation of floral meristems by the apical meristem of racemose inflorescences, such as that of *Arabidopsis thaliana*. The location of the youngest floral meristems is determined, as well as the calculation of the auxin concentration in the vicinity of the apex in agreement with measurements obtained using new molecular methods (Galvan-Ampudia et al. 2020) (see Chapter 5).

Our analysis of a pentamerous *Arabidopsis thaliana* mutant shows the correlation between the location and identity of perianth organs.

We were also able to model the post-merismatic development of the *Illicium anisatum* flower, which has a spirally initiated gynoecium that becomes whorled at anthesis.

Bull-Hereñu et al. (2022) observed that in many Fabaceae, sepal initiation is unidirectional from the abaxial (K1; K = Kelchblatt in German, not to be confused with S = stamen) to the median (K2, K3) and then the adaxial side (K4, K5). Modeling of the forces acting on these floral meristems shows that the pressure of the bracts located above a given flower within the racemose meristem reinforces the inhibition generated by the two ephemeral bracteoles of that flower, thus being responsible for this particular developmental sequence (see Chapter 5).

Some Magnolia species have Fibonacci phyllotaxis (e.g. *M. solangeana*), particularly if the vegetative spiral extends into the flower. But in other species (e.g. *M. denudata*), the stamens are too small compared to the sepals to support this type of phyllotaxis. The numbers of left and right spirals in their androecium are equal, wrongly suggesting that phyllotaxis is complex whorled (see Chapter 6).

Basal angiosperms (the ANA grade) have either perfectly spiral (*Amborella*, *Illicium*) or perfectly whorled (*Cabomba*, *Victoria*) flowers. In the former, the number of organs of each type is very labile (*Illicium* has 7--33 tepals, 4--50 stamens and 5--21 carpels), whereas in whorled flowers, numbers are perfectly determined (six tepals, six stamens and three carpels in *Cabomba*). The perianth of some Magnoliales (*Liriodendron*) and Laurales (*Daphnandra*) presents the combined advantages of spiral and whorled phyllotaxis: packing optimization combined with a consistency between organ position and identity.

Organ initiation in Fibonacci phyllotaxis takes place at the periphery of the central zone or "front" of the meristem. The number of organs formed in one turn of this front belongs to the Fibonacci sequence: 3, 5, 8, etc. In *Liriodendron* or *Daphnandra*, successive fronts give rise, during floral development, to pseudo-whorls, each cycle keeping track of its spiral origin (see Chapter 6). Thus, later in evolution, the quincuncial aestivations will be formed by this process.

Some whorls contain twice as many organs as their predecessors ("dédoublement"), as de Candolle (1813) had already noticed. For instance, while the three perianth whorls of *Anaxagorea crassipetala* are trimerous, the first stamen whorl is hexameric. According to our model, if the sepals of the third whorl are relatively small, then the auxin field of the sepals of the second whorl still influences the initiation of the first stamens, causing their doubling.

We explain "mechanically" the influence of the bracteoles on sepal arrangement, validating the intuition of Eichler (1875) who had searched in vain for an explanation for his observations (see Chapter 7).

We show that in the absence of bracteoles, within racemes such as that of *Antirrhinum majus*, the forces shaping the perianth are not due to biochemical activity, but to contact pressure from the bracts subtending the flowers above each developing flower in the inflorescence. Similarly, within a cyme, if the first-order lateral flowers lack bracteoles, then the arrangement of their calyx is due to pressure from their own bract and from the terminal flower (e.g. *Caltha palustris*).

Eichler also highlighted the relationship between the development of cymes and the identity of the bracteole in the axil of which each new flower is initiated. We developed a geometric method to reconstruct inflorescences from their building blocks, that is, floral meristems. Thus, the difference between a scorpioid cyme (*cincinnus*) and a helicoid cyme lies in the position of the bracteole of a flower in the axil in which the next flower develops. With our method, we reconstructed the *cincinnus* of *Petunia* (corresponding to the SEM of Castel et al. (2010)), *Salvia verticillata* (from the SEM of Claßen-Bockhoff and Bull-Hereñu (2013)) and Eichler's inflorescence diagrams (1875, 1878). The complex *cincinnus* of *Phenakospermum guyannense* requires two bracteoles per flower, even though the plant is a monocot (Kirchoff 2003)!

Inflorescences surrounded by involucral leaves or spathes highlight the effect of the pressure exerted by these organs on the floral meristems they enclose (e.g. *Salicornia*, *Frankenia*, *Mirabilis violacea*) (see Chapter 7). Thus, our model of the phyllotaxic engine developed in Chapter 4 integrates both biochemical "forces" (inhibition due to auxin) and mechanical forces (contact pressure).

The work of Levitov (1991), based on a high-level mathematical method (the theory of symmetry groups, of which our model (trajectories of meristems in phase space; see Chapter 5) is an example), showed that "phyllotaxic" spirals, in particular of the Fibonacci type, are not the prerogative of plants alone, but are also found in several purely physical phenomena (e.g. Rayleigh-Bénard convection, flux lattices in layered superconductors, drops of ferrofluid in a magnetic field). In plants, these spirals are only the application of a general property of symmetry whose advantages have been exploited by natural selection.

1

An Introduction to Phyllotaxis

1.1. A game of spirals

Bud cross-sections allow us to precisely determine the position of leaf *primordia*, that is, leaves at early stages of development. We can see that in many cases, such as that of *Pinus pinea*, the arrangement of these primordia form systems of spirals called *parastichies* that are visible on a *centric representation* (Figure 1.1).

Here, five left-turning and eight right-turning parastichies can be observed. Thus, each primordium belongs to a centric lattice comprising two sets of spirals: we say that *P. pinea* has a (5, 8) spiral *phyllotaxis mode*.

For instance, if we consider primordium number 20 and follow the succeeding primordia (in red in Figure 1.1(C)) in reverse chronological order (20, 21, 22, etc.), the next primordium on the right parastichy to which primordium 20 belongs is primordium 28; 28 – 20 = 8 is equal to the number of right parastichies. The path from primordium 20 to 28 following primordia along their chronological initiation sequence is the *generative* spiral, which is less visible than the parastichies.

For a color version of all the figures in this book, see www.iste.co.uk/walch/phyllotaxis.zip.

To determine the number of parastichies in a spiral structure, we can either count them or subtract the primordium number of two consecutive primordia on the same parastichy.

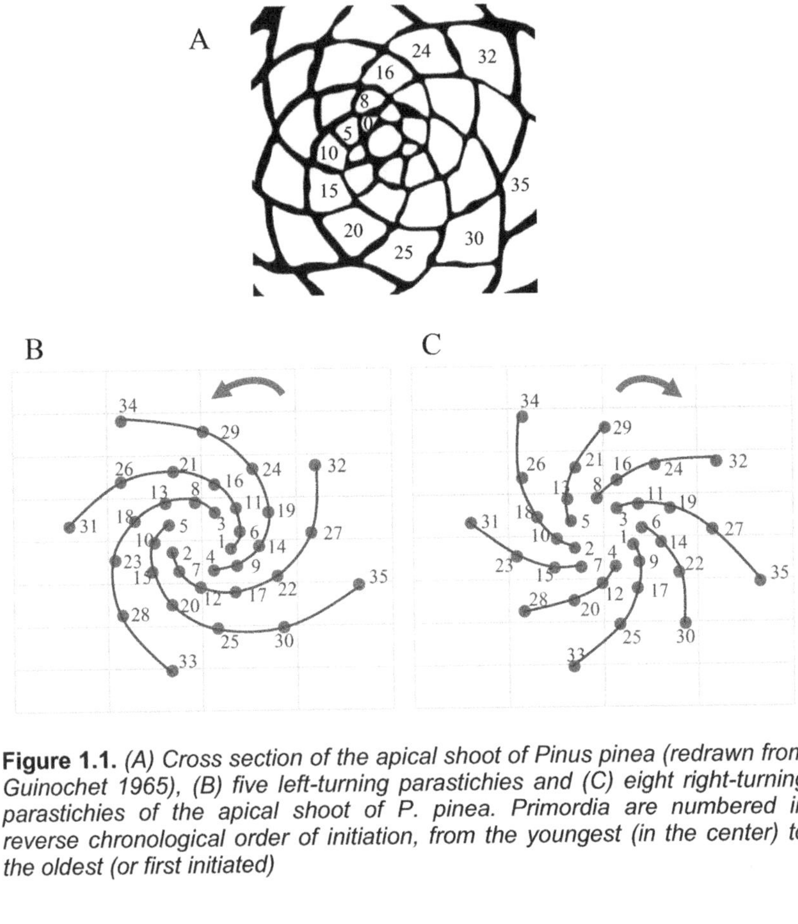

Figure 1.1. *(A) Cross section of the apical shoot of Pinus pinea (redrawn from Guinochet 1965), (B) five left-turning parastichies and (C) eight right-turning parastichies of the apical shoot of P. pinea. Primordia are numbered in reverse chronological order of initiation, from the youngest (in the center) to the oldest (or first initiated)*

Deschatres (1954) made a 2D representation of the stem of *Sedum maximum*. He cut the stem lengthwise, glued it flat on a piece of paper and drew the "lines" linking the different nodes. This is a *cylindrical representation* of a stem (Figure 1.2), as opposed to the centric representations shown in Figure 1.1.

In Figures 1.2(A) and (B), the generative spiral forms line segments with weakly positive slopes that link the primordia in their chronological order, whereas in Figure 1.2(C), parastichies form a lattice of left and right line segments. The difference between the numbers of two consecutive points on a same parastichy is, respectively, 2 and 3 (e.g. 15 – 13 = 2 and 18 – 15 = 3). Thus, this part of the stem of *Sedum maximum* has a (2, 3) phyllotaxis mode.

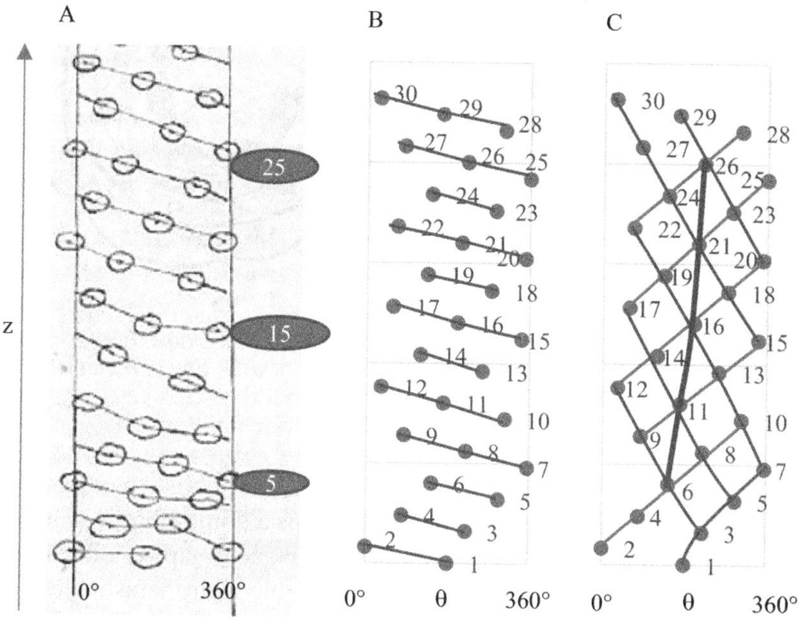

Figure 1.2. *Cylindrical representations of the stem of Sedum maximum. (A) Drawing from Deschatres (1954). (B) Generative spiral. (C) Parastichies and one orthostichy (in red). On the x-axis: angles from 0° to 360° measured on a circle around the stem; on the z-axis: the height along the stem*

We can also see near-vertical lines, such as those on which are found 6-11-16-21-26 (in red) or 4-9-14-19-24-29 in Figure 1.2(C), which are medians of the two parastichies; the difference in the chronological number of two primordia on these lines is equal to the sum of the number of parastichies: 11 – 6 = 5 = 2 + 3. These lines would be called *orthostichies* if they were vertical but in this case are called *pseudo-orthostichies* because they are tilted.

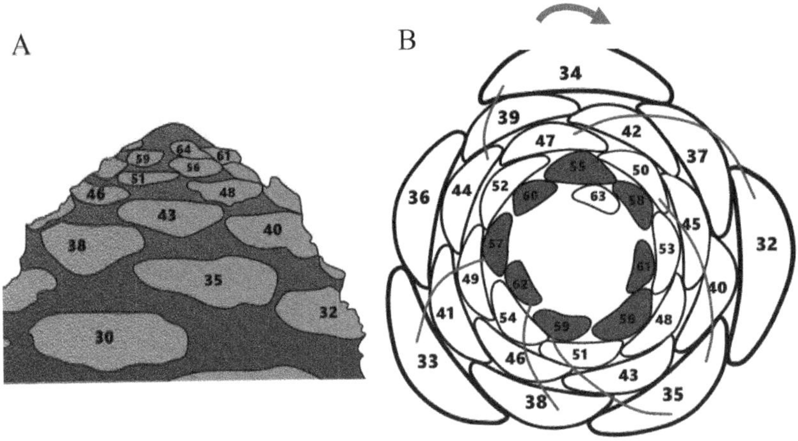

Figure 1.3. *(A) The meristem of Linum usitatissimum (redrawn from Meicenheimer 2006). Numbers correspond to the chronological order of the leaf primordium on the shoot. (B) Cross-section of a 22-day old shoot apical meristem (from Williams 1975)*

Primordia arise near the *apex* (or tip) of the meristem and move outward and down the meristem, which forms a dome (Figure 1.3(A)). On the cross-section of a 22-day old stem apex of *Linum usitatissumum* (Figure 1.3(B)), we can see the right parastichies (or clockwise, e.g. 44-39-34) and the left one (or anti-clockwise, e.g. 51-43-35). Thus, this phyllotaxis mode is (5,8).

Primordia (e.g. number 63 in Figure 1.3(B) and 12 in Figure 1.4(A)) appear at regular intervals on a crown, or *front* (in black on Figure 1.3B and Figure 1.4A), located at a given distance from the meristem apex (Hotton et al. 2006; Golé et al. 2016). The time interval between the initiation of two consecutive primordia is called the *plastochron* (T). Richards (1951) defined the *plastochron ratio* (r_i) as the distance from the apex of primodium i (R_i) over the distance of its chronological successor (R_{i+1}):

$$r_i = \frac{R_i}{R_{i+1}} > 1 \qquad [1.1]$$

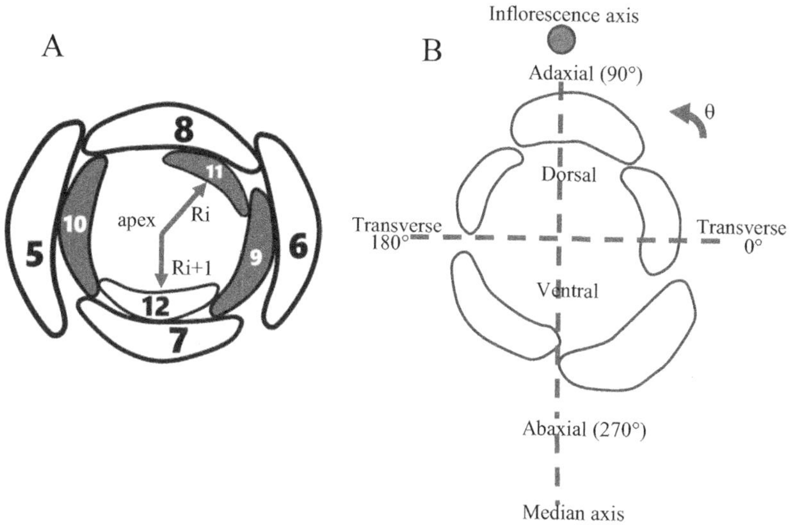

Figure 1.4. *(A) The crown where new primordia arise (number 9, 10, 11 in black and 12 the youngest one).* R_i *and* R_{i+1} *are used to calculate the plastochron ratio. (B) Generic floral diagram (see Ronse de Craene 2022), with names of flower part locations: transverse and median axis; adaxial/abaxial side of the median axis; dorsal/ventral parts ot the flower. The angles on the trigonometric circle (*θ*) are given*

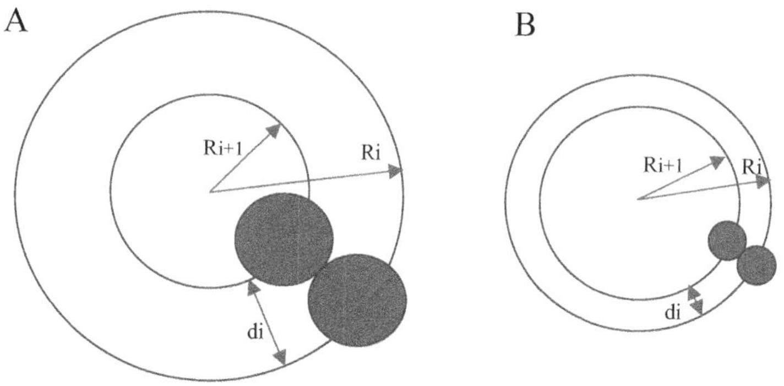

Figure 1.5. *The relationship between plastochron ratio (*R_i/R_{i+1}*) and organ size (di). (A) Large plastochron ratio. (B) Small plastochron ratio*

For instance, if the distance from the apex of primordium 11 is equal to 77 µm and that of primordium 12 is 70 µm (values measured in *Linum usiratissimum* from Williams (1975) drawing), then the plastochron ratio when primordium 12 appears is 77/70 = 1.1.

Since organs are in contact, high plastochron ratios are equivalent to organs being large (Figure 1.5(A)), whereas small platochron ratios (close to 1) are equivalent to organ being small (Figure 1.5(B)).

1.2. Fibonacci phyllotaxis

Williams (1975) represented the cross-section of the shoot apical meristem of *Linum usitatissimum* in 1-, 4-, 8-, 11-, 15-, 18-, 22- and 50-day old buds and precisely measured the evolution of the plastochron ratio with age: this growth rate was found to decrease over time (Figures 1.6(A) and (B)).

Age (days)	Primordium number	Plastochron ratio	Age (days)	Primordium number	Plastochron ratio
1	Cot.–2	1.330	25	63–73	1.045
4	2–4	1.189	29	84–94	1.041
4	4–6	1.152	32	100–110	1.044
8	4–14	1.101	36	125–135	1.035
11	10–20	1.082	39	151–161	1.030
15	20–30	1.073	43	173–183	1.033
18	35–45	1.052	50	202–212	1.034
22	51–61	1.049			

Cot. = cotyledons.

Table 1.1. *Evolution of the plastochron ratio with primodium age in Linum usitatissimum (from Williams 1975)*

As the growth rate of the meristem measured by plastochron ratio decreases, more primordia are found at the front and the number of parastichies increases (Figures 1.3 and 1.6). Thus, on day 22, the plastochron ratio has decreased to 1.049 and the phyllotaxis mode increased from (3,5) to (5,8).

The number of parastichies takes on the following values as the plant grows: 2, 3, 5, 8, 13, etc., where each number is the sum of the two preceding numbers: this is the *Fibonacci sequence* whereby the ratio of consecutive terms, such as 3/2, 5/3, 8/5, 13/8, etc., get closer and closer to the *golden ratio* (1.618).

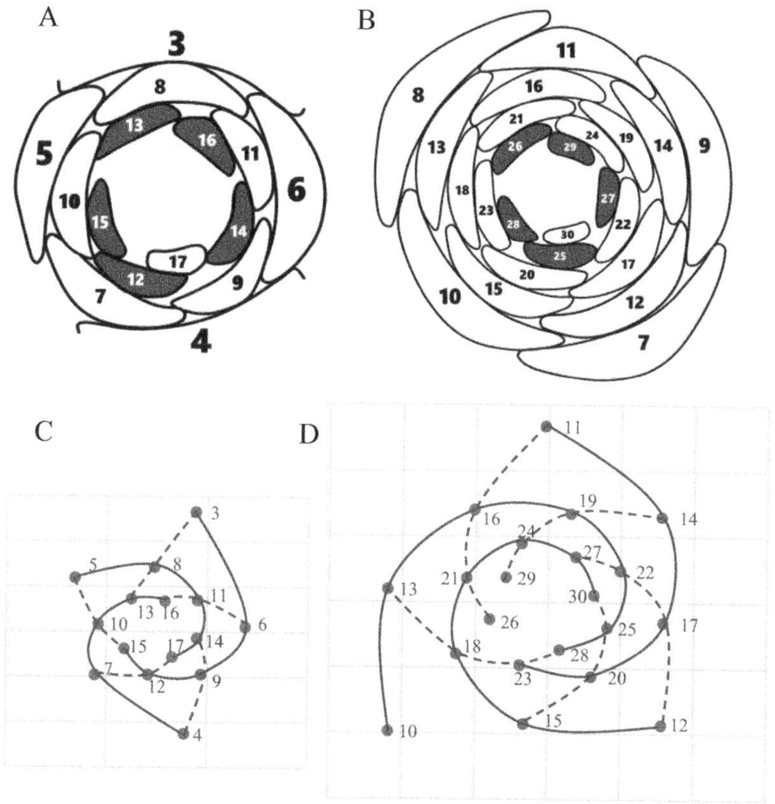

Figure 1.6. *Representation of an 11-day old (A and C) and 15-day old (B and D) vegetative meristem of Linum usitatissimum from Williams (1975) and model developped in Chapter 4. Phyllotaxis mode is (3, 5)*

The angle between two chronologically consecutive primordia is called *divergence angle* (Figure 1.7). In the same way as the ratio of Fibonacci numbers converges toward the golden ratio, the divergence angles in meristems that have numbers of parastichies that are terms of the Fibonacci sequence are close to the golden angle, which is equal to (see Chapter 3):

$$\varphi = \frac{3-\sqrt{5}}{2} \times 360° \approx 137.51° \quad [1.2]$$

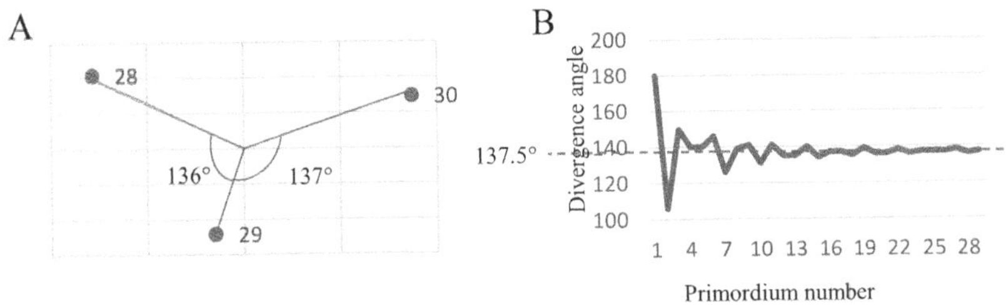

Figure 1.7. *(A) The divergence angle between primordia 19 and 20, and primordia 20 and 21 in Linum usitatissimum. (B) The divergence angle between two consecutive primordia converges to the golden angle (137.5°)*

In *L. usitatissimum*, the divergence angle, which is initially 180° between the first two opposite primordia (cotyledons), oscillates between 130° and 140° after primordia 13, and only between 137 ± 3° from primordia 22 onwards (Figure 1.7(B)).

1.3. Lucas phyllotaxis

In the vast majority of plants with spiral phyllotaxis, the number of parastichies follows the terms of the Fibonacci sequence. Some plants, however, are arranged according to other sequences, and the most common of these is the Lucas sequence.

Rutishauser (1998) drew the cross-section of *Sedum sexangulare*, which has the particularity of having 18 primordia in a (3, 4) phyllotaxis, corresponding to the *Lucas sequence* (2, 1, 3, 4, 7, 11, etc.). The divergence angle is approximately 101° and the plastochron ratio is 1.12 (both are average values for the first 18 primordia).

S. sexangulare got its name because most members of this species have six orthostichies; however, some have not six but seven orthostichies (Deschatres 1954). This is the case for the specimen examined by Rutishauser (1998) (Figure 1.8).

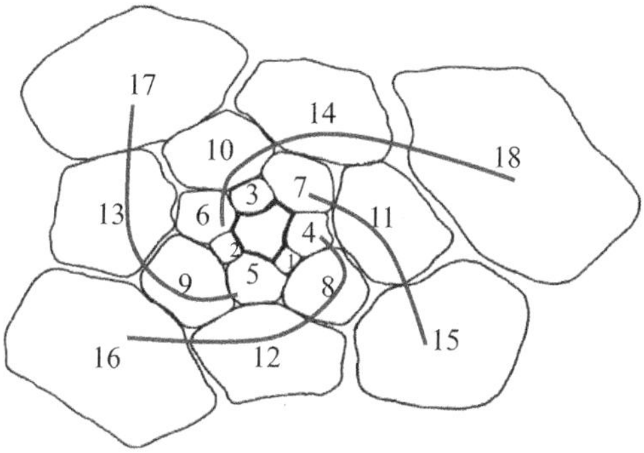

Figure 1.8. *The leaf apical meristem of Sedum sexangulare examined by Rutishauser (1998). The phyllotaxis mode is (3, 4). Only the four right parastichies are shown. Primordia are numbered in reverse chronological order, from the youngest (in the center) to the oldest*

The divergence angle between succesive primordia in meristems whose numbers of parastichies are terms of the Lucas sequence is close to the Lucas angle, which is equal to:

$$\frac{5-\sqrt{5}}{10} \times 360° \approx 99.501° \qquad [1.3]$$

The evolution of Lucas phyllotaxis as plastochron ratios decrease over time is as follows: (3, 4) -> (4, 7) -> (7, 11). In *S. sexangulare*, the divergence angles range between 90° and 103° (Rutishauer 1998).

1.4. Whorled phyllotaxis

Spiral arrangement is not the only type of phyllotaxis. Whorled patterns consist of alternating *whorls*, each composed of the same number of organs (*merosity*). The divergence angles between organs within a whorl are constant and equal to 360° divided by the number of organs. It is 180° = 360°/2 for a dimerous whorl, 120° = 360°/3 for a trimerous and 90° = 360°/4 for a tetramerous whorl. Whorled patterns present perfectly vertical orthostichies (Figure 1.10(B)) on the shoot unlike pseudo-orthostichies of spiral phyllotaxis. They have the same number of clockwise and anti-clockwise parastichies (Figure 1.10(A)).

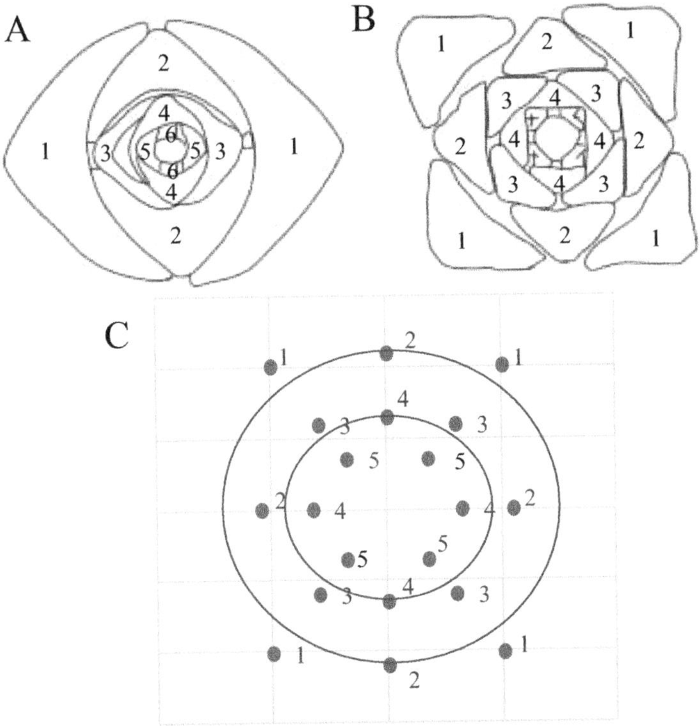

Figure 1.9. *(A) Cross-sections of the apical vegetative meristem of Honkenya peploides (Caryophyllaceae) and (B) Plantago webbii (Plantaginaceae) from Rutishauser (1998). (C) Concentric whorls of P. webbii*

In a whorled arrangement, organs are located on concentric circles (Figure 1.9(C)). On the same whorl, the primordia have the same number because they are indistinguishable and they are located at the same distance from the apex, so their plastochron ratio is equal to 1:

$$\frac{R_{i+1}}{R_i} = \frac{R_i}{R_i} = 1 \qquad [1.4]$$

Between two whorls, the diameter of the circles is not the same, so plastochron ratio between these two whorls is superior to 1.

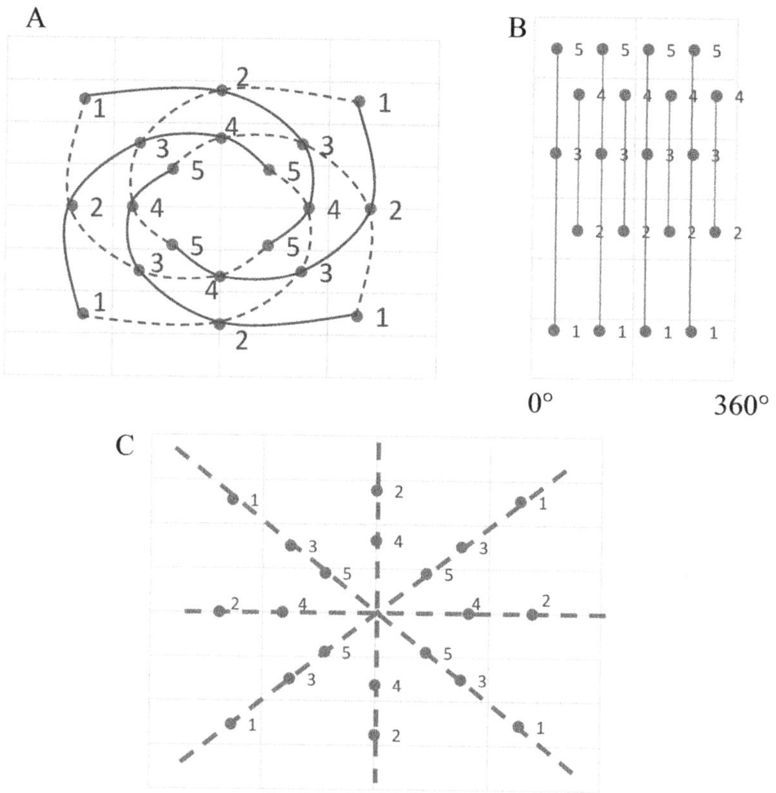

Figure 1.10. *Plantago webbii. (A) Centric representation representing four anti-clockwise (red) and four clockwise (dashed lines) parastichies. (B) Cylindrical representation showing orthostichies. (C) Centric representation with axes of symmetry*

The vegetative meristem of *Honkenya peploides* (Caryophyllaceae) is *distichous* (Figure 1.9A). Pairs of leaves are initiated simultaneously at the periphery of the apical meristem, with the leaves opposite each other. The vegetative meristem of *Plantago webbii* (Plantaginaceae) is tetramerous. At the shoot apex, the four members of a whorl are initiated simultaneously (Figure 1.9(B)) (Rutishauser

1998). Each whorl has the same number of primordia. The phyllotaxis mode is (4, 4; Figure 1.10(A)).

In spiral phyllotaxis, the divergence angle is an *irrational number*, that is, calculated using a square root. In whorled arrangements, the divergence angle is obtained as a simple fraction of the trigonometric circle: it is a *rational number* such as in *Plantago webbii*:

$$\varphi = \frac{1}{4} \times 360° = 90° \qquad [1.5]$$

The vegetative meristem of *Plantago webbii* shows four axes of symmetry. It is *polysymmetric* or *actinomorphic* (radially symmetrical) according to the vocabulary of botanists (Figure 1.10(C)). Spiral phyllotaxis is *asymmetric* (Figure 1.1).

2

A History of Theoretical Phyllotaxis

In *Recherches sur l'Usage des Feuilles dans les Plantes* (1754), the Genevan naturalist Charles Bonnet dedicated a chapter on the "arrangement of leaves on a stem or a branch", continuing the work of François Boissier de Sauvages, professor of medicine at the university of Montpellier, and the famous Swedish naturalist and father of binomial nomenclature Carl Linnaeus. Bonnet distinguished five types of leaf arrangements:

– "alternes" (opposite): leaves are arranged along the stem in two vertical rows that are parallel to the stem and diametrically opposite each other;

– "à paires croisées" (*decussate*): leaves are arranged in opposite pairs, and successive pairs are perpendicular to each other;

– "à feuilles verticillées" (whorled): like the spokes of a wheel around a hub;

– "en quinconces" (*quincuncial*): leaves are arranged in groups of five, the sixth leaf being located on the same vertical line as the first; as these leaves are sufficiently distant from each other, this does not affect their function;

– "en spirales redoublées" (multi-spirals): leaves are arranged along multiple parallel spirals. These spirals can be in threes (pine) or fives (fir).

For a color version of all the figures in this book, see www.iste.co.uk/walch/phyllotaxis.zip.

The German naturalist Wilhelm Schimper was the first to analyze spiral arrangements from a mathematical point of view (1830). He introduced the concepts of generative spiral, divergence angle and secondary spiral (parastichy in the modern sense). He supposed that all divergence angles were fractions of the circle (i.e. rational numbers) and noticed that these could generally be expressed by dividing a term of the Fibonacci sequence (1, 1, 2, 3, 5, 8, 13, etc.) by the second next term in the sequence, that is, 1/3 = 360°/3 = 120°, 2/5 = 2 × 360°/5 = 144°, 3/8 = 3 × 360°/8 = 135°, 5/13 = 5 × 360°/8 = 138.5°, 8/21 = 8 × 360°/21 = 137.1°, etc. The Bavarian botanist Alexander Braun (1835) noted that the number of right and left parastichies were very often consecutive terms of the Fibonacci sequence (Adler et al. 1997).

For Schimper and Braun, the different angles obtained by the division of consecutive terms of the Fibonacci sequence "actually occur in nature".

Auguste Bravais, an École Polytechnique alumnus known for his work in crystallography (1850), and his brother Louis Bravais, a physician and botanist, disagreed with the theory of Schimper and Braun (1837a): they postulated that the divergence angle between leaves is the same for all plants and is comprised between 137° and 138°. More precisely, it is obtained by the limit of the ratio of the terms of the Fibonacci sequence 1/3, 2/5, 3/8, etc., an irrational number that cannot be obtained by simple division, but only as a limit and which in degrees is 137°38′28″ (see Chapter 3). This angle divides a circle (360°) in "extreme and mean ratio", which is to say by the "golden ratio", which the Bravais brothers refrain from naming directly.

The divergence angle is the angle that separates a primordium from its chronological successor. It may seem hard to measure, but on the bud of *Pinus pinea* (Figure 1.1), we can see that primordia number 6 and 27 are located approximately on the same radius. Thus, the angle which separates them is close to a multiple of 360°. By following the primordia chronologically, we can see that eight turns are required to pass through 21 primordia (21 = 27 – 6), so that the divergence angle is roughly equal to 8 × 360/21 = 137.14°.

The different values obtained by Schimper and Braun appear to be due to measurement issues, according to Bravais and Bravais. Indeed, on an average-sized stem, say 1 cm in diameter, a 1° arc is less than 8 hundredth of a millimeter, a length which cannot be seen with the naked eye.

Nevertheless, Bravais and Bravais recognized that depending on the length of the interval between two consecutive nodes along the stem (the merithal or internode), the number of parastichies (or "secondary spirals") increases from 2, if the internode is long, to 3, 5, 8 or 13 and can reach 21, 34 or even 55 for the Synanthérées (nowadays, Asteraceae) if the internode is short.

For Bravais and Bravais, the generative spiral remains the same throughout the insertion of the leaves on the stem and continues with the leaflets of the involucrum of Asteraceae inflorescences, then the bracts, which would be the consequence of having a constant divergence angle, whereas for Schimper and Braun, there would be a succession of spirals with different divergence angles. Bravais and Bravais list flowers whose organs are on the same generative spiral as the leaves: *Trollius*, *Hepatica*, *Illicium floridanum*, *Camellia*, *Polygonum*, etc. In the same way, the generative spiral is continued in underground stems such as the bulbs of Liliaceae, the rhyzomes of ferns, rose bushes, *Belladonna*, meadow saxifrage and potato, which bears tubers at the tip of underground branches.

Some plants are organized following sequences that are different from that of Fibonacci, such as 1, 3, 4, 7, 11 and so on (the Lucas sequence), as in *Sedum reflexum* and *Dipsacus sylvestris*, which have 18, 29 and 47 parastichies. Their corresponding divergence angle converges toward 99°39′.

Louis Bravais and Auguste Bravais examined the surface of a cylindrical stem by cutting it lengthwise and placing it on a flat surface: thus, they represented phyllotaxis as a cylindrical lattice.

In 1868, the German professor Wilhelm Hofmeister hypothesized that primordia initiate at regular time intervals (T = plastochron) at the meristem apex in the largest gap left available by preceding primordia. This idea meant that the study of phyllotaxis should be carried out on primordia in the bud rather than by examining leaf arrangement on a mature stem.

In 1878, the Swiss botanist Simon Schwendener published a comprehensive treatise on phyllotaxis. He suggested that the arrangement of leaves is the consequence of the pressure exerted by each primordium on neighboring primordia. He introduced the (m, n) notation, as in for instance the phyllotaxis mode (5, 8) with m parastichies turning in one direction and n parastichies in the other.

The British botanist Arthur Church, who taught at Oxford University, published in 1904 numerous photographs and drawings made with a camera lucida, illustrating the different types of phyllotaxis. We owe him the centric representation of the meristem, which is complementary to the cylindrical lattice of Bravais and Bravais. He introduced the concept of bulk ratio, that is, the ratio of the radius of the meristem to that of the youngest primordia, which would also follow the Fibonacci sequence.

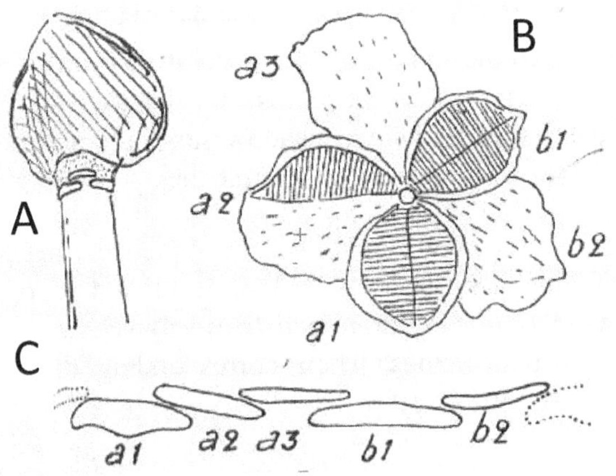

Figure 2.1. *(A) The calyx of Anemone japonica, from Plantefol (1949); (B) underside view, (C) cross-section, to be compared with a quincuncial arrangement*

A "French" school of phyllotaxis was developed by Plantefol (1949) who taught botany at La Sorbonne.

> The old theories, founded on the existence of a "generative spiral", are greatly insufficient. Mechanical theories, the type being that of Hofmeister, seem to work for certain types of stems but in the case of others, are "confronted with irreducible facts" (Deschatres 1954; a disciple of Plantefol).

According to Plantefol, leaves and flower organs arise in a contiguous manner along one or more "leaf helices" turning in the same direction. Thus, sepals K1 (= a1), K3 (= a2) and K5 (= a3) of *Anemone japonica* would be formed along one helix, and K2 (= b1) and K4 (= b2) along a second helix (Figure 2.1(C)). Leaf helices were a dead end.

In 1951, F. J. Richards of Imperial College London introduced a notion that would become fundamental: the plastochron ratio, that is, the ratio of the distances from the center of the meristem of two successive primordia. A second parameter was needed to fully describe the position of all primordia on the meristem: the angular position (θ). This allowed the use of a system of polar coordinates (R, θ).

Richard also defined a "*phyllotaxis index*" involving the double decimal logarithm of the plastochron ratio, which he noted was a linear decreasing function of the plastochron ratio (see Chapter 3).

Snow and Snow (1952) from Magdalen College, Oxford, generalized Hofmeister's approach by removing the imposed periodicity of the process and postulating that new primordia form as soon as there is enough space for them within the apical meristem.

The work of the Dutch botanist Gerrit van Iterson (1907) remained unnoticed for more than half a century, probably because it was too mathematical. It was rediscovered by Ralph Erickson (1973) from the University of Pennsylvania. His method consisted in stacking disks of equal diameter in a cylinder (Figure 2.2(A)). From this, we can observe parastichies (straight lines) corresponding to Fibonacci spirals. The system goes up the Fibonacci sequence as the *diameter of the disks* decreases. The divergence angle is the second parameter of the model: it is the angle between two consecutive disks, that is, the horizontal distance on the lattice between these two disks (the total width of the cylinder being 360°).

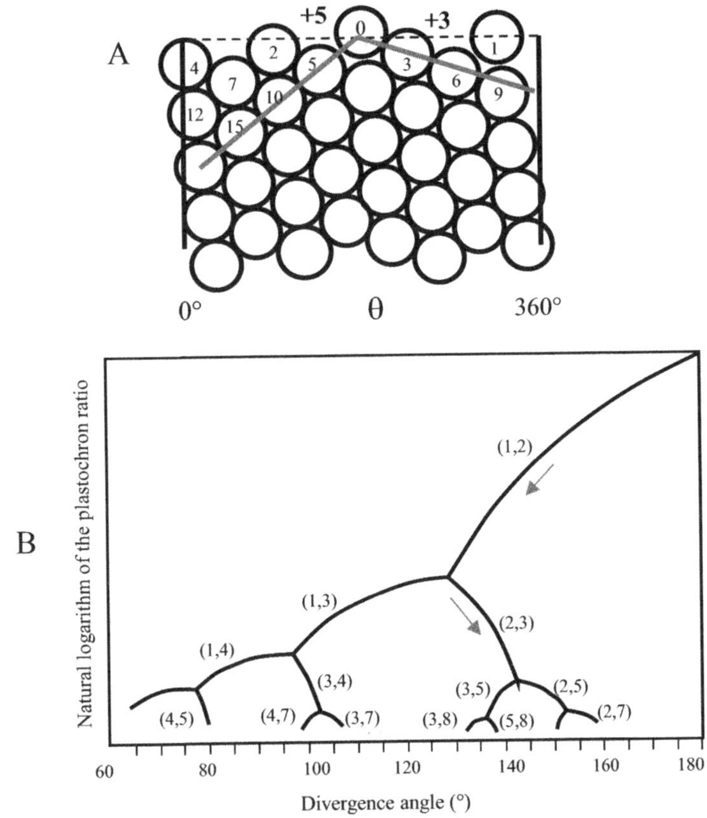

Figure 2.2. *(A) Packing a cylinder with disks of equal diameter: parastichies (blue straight lines) are here in mode (3, 5). (B) The van Iterson diagram (from Erickson 1973)*

A diagram in phase space (divergence angle, natural logarithm of the plastochron ratio) shows the phyllotaxis modes: this is the famous *van Iterson diagram* (Figure 2.2(B)). Godin et al. (2020) interpreted plant development as "traveling" down the van Iterson diagram.

Monocots and dicots start on a branch with a low mode, for instance, (1, 1) with the second primordium opposite the first. Then, the plastochron ratio decreases, which leads to either a (1, 2) or a (2, 1) pattern depending on the *chirality* of the spiral (turning right or left). If the (1, 2) branch is chosen, the "Fibonacci rule" imposes the phyllotaxis mode (2, 3). The decrease in disk diameter inexorably leads to higher Fibonacci modes (3, 5 → 5, 8, etc.).

In 1977, Graeme Mitchison from the Laboratory of Molecular Biology at Cambridge University (UK) carried out computer simulations of the inhibitory mechanism suggested by Schoute (1913), that is, primordia produce an inhibitor that prevents the formation of new leaves in their vicinity. This inhibitor diffuses from its source so that a new leaf can only arise at a certain distance where the concentration of the inhibitor falls below a certain threshold. The inhibitor concentration is the sum of the concentrations produced by pre-existing leaves and decreases exponentially with the distance from the source.

Veen and Lindenmayer (1977) from the Netherlands modeled the diffusion mechanism of an inhibitor that could be responsible for the generation of phyllotaxis patterns using the equations of Fick's law (a physical law that models the diffusion process). Their model reproduced the van Iterson diagram.

Marzec and Kappraff (New York) mathematically studied the effect of divergence angles on leaf distribution (1983). They showed that certain remarkable angles optimize this distribution, meaning that they produce the densest meristems: they can all be expressed as a function of the golden ratio:

$$\tau = \frac{1+\sqrt{5}}{2} \qquad [2.1]$$

These angles are therefore all irrational, and we find among them the golden angle (137.507°) as well as the limit of the Lucas sequence (99.501°; see Chapter 3).

The seminal papers of modern mathematical phyllotaxis were written by two scientists at the Laboratoire de Physique Statistique of the École Normale Supérieure, Paris: Stéphane Douady and Yves Couder. Their first paper published in 1992 went unnoticed, probably because it was published in a physics journal.

The principles of their model are as follows (Douady and Couder 1996a):

1) Primordia form at the periphery of the apex (of radius R_0) and as the meristem grows, they move away from the apex at a radial velocity V(R), an exponential function of the distance from the center (R).

2) New primordia form regularly over time, the amount of time separating the initiation of two consecutive primordia being the plastochron T. It is an iterative process.

3) New primordia form in the largest possible space left by pre-existing primordia since they are subjected to repulsive forces generated by the latter.

4) Different repulsion potentials have been used (the inverse function of the distance between a given point where the potential is calculated and the generative primordium; the decreasing exponential function of this distance, etc.).

If R_0 is the radius of the central zone, V_0 the centrifugal radial velocity and T the delay between the appearance of two primordia, then V_0T is the distance covered by a primordium during one plastochron. The growth index $G = V_0T/R_0$ is a dimensionless number that characterizes the growth of the meristem irrespective of size and time scale.

If growth is exponential, then the distance from the center of a primordium is:

$$R = R_0 \exp(V_0 t/R_0) = R_0 \exp(t/TG)$$

$$\text{hence: } r = \frac{R(t+T)}{R(t)} = \frac{e^{V0(t+T)/R0}}{e^{V0t/R0}} = \exp(V_0T/R_0) = \exp(G)$$

and G= ln(r) is the growth index equal to the natural logarithm of the plastochron ratio.

Box 2.1. *Growth index (Douady and Couder 1996a)*

For each simulation, the *growth index* G decreases with each iteration. For example, starting with G = 1 and a (1, 1) mode (distichous) and ending with G = 0.01, they obtain a divergence angle of 137.47° and (13, 21) parastichies (Fibonacci phyllotaxis). Starting with G = 0.3, the divergence angle is 99.49° and the final Lucas mode is (11, 18). The iterations reproduce the van Iterson diagram.

In a second model (Douady and Couder 1996b), instead of specifying the plastochron T, Douady and Couder allow the initiation of a new primordium at a point at the periphery of the apex where the repulsion energy produced by pre-existing primordia falls below a minimum *threshold* U_0. In fact, they replace this condition by an equivalent one: the distance between the new primordium and its neighbors must be above a threshold d_0. The amount of time for a new primordium to appear is now no longer given a priori, and furthermore, the meristem is modeled as a cone. The system generates its own primordium initiation periodicity. Unlike the first model, where primordia only appear sequentially, primordia can appear simultaneously and thus form whorls. Depending on the shape of the cone, the model can converge toward whorled phyllotaxis (2, 2), (3, 3), etc. or generate spirals.

In a third article, Douady and Couder (1996c) studied the transition from one phyllotaxis mode to another: distichous to spiral and vice versa. Snow and Snow (1935) had shown that by slitting the apex of *Epilobium hirsutum*, its phyllotaxis, which is normally decussate, becomes temporarily spiral and only much later becomes decussate again. Under certain parameters, this third model simulates the same pattern. Whorled and spiral phyllotaxis can therefore be produced by the same mechanism, the only difference being the initial conditions. The principal factor that determines the most stable phyllotaxis is the shape of the apex, the stability of the decussate mode being highly increased by the effect of conicity.

According to Douady and Couder (1996a), we must not think that the functional properties of the convergence toward an irrational angle explain the prevalence of Fibonacci phyllotaxis. All stems are not vertical and the sun is not always at the zenith. Besides, how can we explain the high frequency of decussate phyllotaxis? These interpretations may be viewed as finalistic, and if selection pressures can favor certain arrangements, the mechanism underlying the formation of these structures must exist before any selection pressure can act on it.

When these physicists came to present their work to botanical researchers, they were regarded as extra-terrestrials. The fact the "phyllotaxis" had been reproduced in a laboratory from a few droplets of ferrofluid in a magnetic field is not something that was going to reassure naturalists!

For his PhD at the University of Calgary, Richard Smith, supervised by Prsemyslaw Prusinkiewicz and the Swiss botanist Cris Kuhlemeier, investigated phyllotaxis using computer simulation models (Smith et al. 2006). He modeled the meristem in 3D using a mathematical tool much used by computer scientists, B-spline functions. In their model, the "inhibition function" produced at a point on the meristem is inversely proportional to the distance from the primordium and decreases exponentially with primordium age. At each iteration, the algorithm calculates this function at all points of the meristem taking into account the position of all pre-existing primordia and adding up the inhibition functions produced by these primordia. If at one point on the front the value of the field is below a given inhibition threshold, then a new primordium appears at that point. The process can then be repeated.

For certain threshold values, the model generates a Fibonacci spiral whose divergence angles correspond to those that have been measured in the first 10 leaf primordia of *Arabidopsis thaliana*. A higher threshold can generate a Lucas spiral.

Having a single inhibitor per primordium does not generate phyllotaxis where multiple primordia appear simultaneously as found in whorled arrangements. The second model of Smith et al. (2006) therefore has a second inhibitor, whose effect is not instantaneous.

This model can generate most phyllotaxis observed in nature. A transition from decussate to spiral phyllotaxis can also be generated.

Kitazawa and Fujimoto (2015) from Osaka University modeled the floral phyllotaxis of *Silene coeli-rosa* (Caryophyllaceae) as it was precisely measured by Lyndon (1978). It is one of the few explicit publications on the evolution of divergence angles and plastochron ratios during floral development in a eudicot. This flower has five sepals and five petals that are initiated in a spiral.

The inhibition potential of the Kitazawa and Fujimoto model decreases exponentially with the distance from the generative primordium and its age (in number of plastochrons).

Kitazawa and Fujimoto also modeled a post-meristematic developmental phase, which takes place after organ initiation and during which phyllotaxis changes in response to the contact pressure exerted by each primordium on neighboring primordia. This stage is modeled by a decreasing exponential function of the distance between all the primordia, in the same way that they modeled inhibition fields and primordia go down the potential gradient where they are located (see Chapter 5).

A different approach was taken by a team of American mathematicians, primarily from Smith College, Massachussetts (Atela and Golé), who reprised the pioneering method of van Iterson (Hotton et al. 2006). They stacked disks in a cylinder and obtained Fibonacci spirals: the smaller the disk, the higher the phyllotaxis mode.

In defense of their method, although it does not consider the molecular and cellular mechanisms that control phyllotaxis, it explains simply how primordia produce the patterns commonly found in plants.

Atela (2011) stacked disks not in a cylinder but in a cone that simulates the decrease in disk diameter because it is wider at the top than at the bottom.

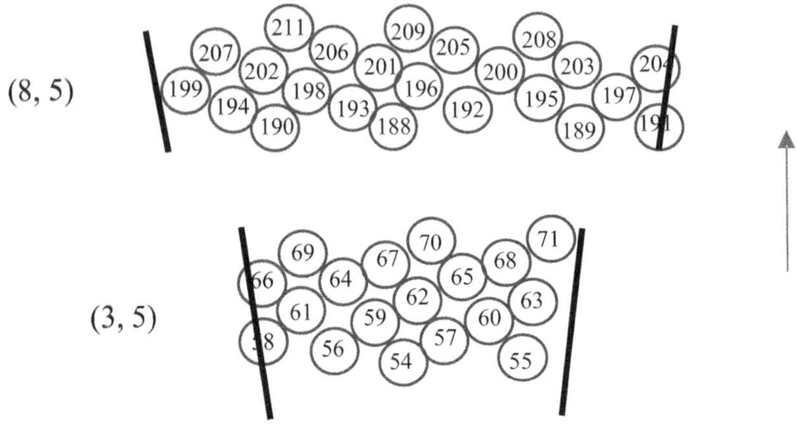

Figure 2.3. *Phyllotaxic transition according to the model of Atela (2011). At the base, the phyllotaxis mode is (3, 5), and increases to (8, 5) and finally (8, 13)*

In 2016, Golé et al. showed that quasi-symmetric phyllotaxis, with nearly equal number of left and right parastichies, arise if the diameter of the disks rapidly decreases compared to those that generate Fibonacci spirals. For geometricians, the plastochron ratio is replaced by the ratio of primordium diameter over that of the meristem (see Chapter 6).

In 2020, Godin et al. published "Phyllotaxis as Geometric Canalization during Plant Development". They hypothesized that geometry, that is, the competition between primordia for space on the meristem, is the cause of Fibonacci spirals. The plastochron ratio is discarded once again in favor of the "geometric ratio", the ratio of primordia diameter over that of the meristem.

3

The Static Model

3.1. Modeling the sunflower capitulum

In sunflowers (*Helianthus annuus*, Asteraceae), leaf phyllotaxis around the stem is initially opposite-decussate, then goes from (2, 3) to (5, 8) and finally (13, 21), whereas the involucrum is in a (34, 55) mode and the capitulum can bear hundreds of florets ((89, 144) mode) (Church 1904). All these numbers belong to the Fibonacci sequence. The sunflower head was simulated by Douady and Couder (1996a) using their dynamical model: plastochron ratio decreased from 2 (cotyledons) to 1.005, then increased back to 1.35. The whole plant shows the continuity of the generative spiral along which, successively, leaves, bracts, ligulae and then florets have formed.

The primary objective of our static model is to simulate the arrangement of organs as it can be observed on a mature plant with only regular decreasing plastochron ratio. Each organ is represented by a disk centered on the point of polar coordinates (R,θ) and whose diameter is equal to the distance to the nearest organ. In this simple model, the divergence angle between organs (ϕ) is constant and the radial distance to the center of capitulum R(n) of the florets is proportional to the square root of the floret number n (Rivier et al. 1984):

For a color version of all the figures in this book, see www.iste.co.uk/walch/phyllotaxis.zip.

$$\theta(n) = n\varphi \quad [3.1]$$

$$R(n) = \sqrt{n} \quad [3.2]$$

In sunflowers, the divergence angle is fixed at 137.51° (the golden angle; equation [3.1]), which produces a Fibonacci spiral, and the law of the square root of the distance (equation [3.2]) is defined so that floret density remains constant from the center of the capitulum to the periphery: the number of florets per unit of surface is independent of the distance from the center of the capitulum.

Floret density (the surface of radius R divided by the number of florets in this surface) is constant:

$$\sigma = \frac{\pi R^2}{n} = \pi$$

By derivation, we obtain for n big:

$$d_n = R_{n+1} - R_n \propto \frac{1}{\sqrt{n}}$$

where d_n is the diameter of floret n.

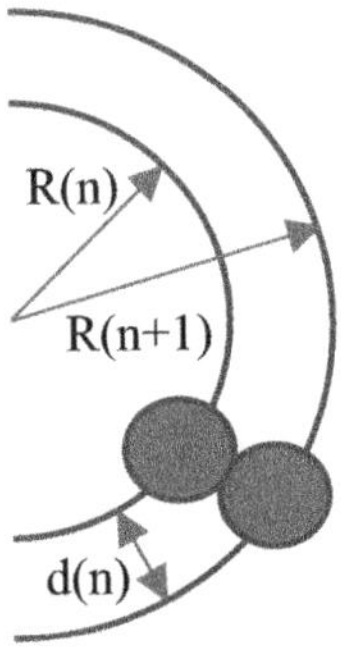

For the plastochron ratio:

$$r - 1 = \frac{R_{n+1} - R_n}{R_n} \propto \frac{1}{n}$$

By Taylor expansion $\ln(1+x) \approx x$:

$$G = \ln(r) \approx \frac{1}{n}$$

or, for n big:

$$\ln(G) \approx -\ln(n)$$

Thus, ln (ln(r)) is a decreasing linear function of the natural logarithm of the organ number (Richard phyllotaxis index).

Box 3.1. *Growth index in the static model*

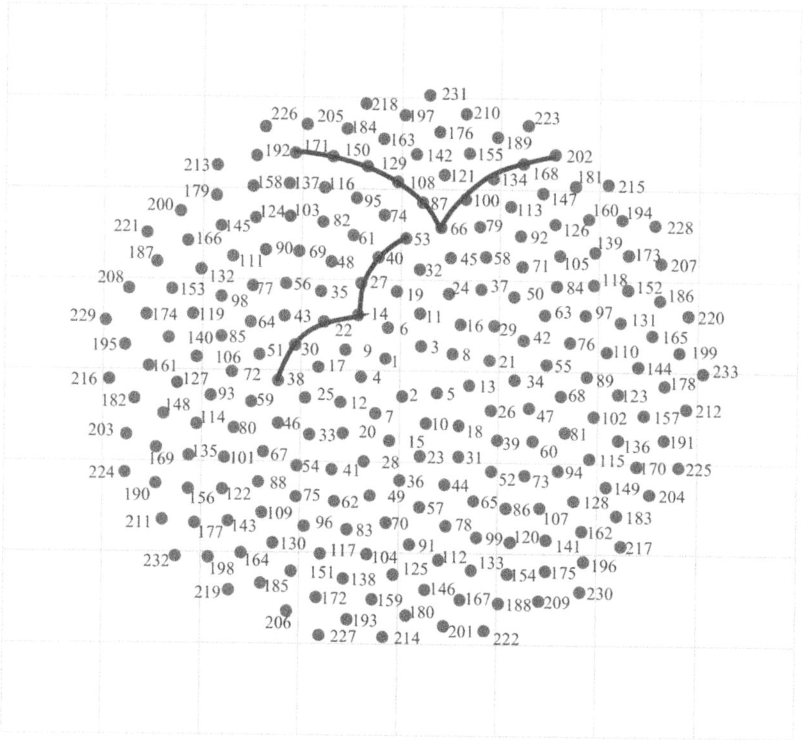

Figure 3.1. *Modeling of a sunflower capitulum according to equations [3.1] and [3.2]*

In this model (equations [3.1] and [3.2]), we can clearly see the parastichies of the "rising phyllotaxis" (Church 1904) of a structure built on the golden angle. Parastichies (14, 27, 40, etc.) and (14, 22, 30, etc.) of mode (13, 8) and parastichies (66, 100, 134, etc.) and (66, 87, 108, etc.) of mode (34, 21) are shown, thus climbing the Fibonacci sequence from the center of the capitulum to the periphery (Figure 3.1).

On a logarithmic scale, ln(G) function of ln(n) is a line with a negative slope. Down this slope, the Fibonacci mode rises regularly (Figure 3.2).

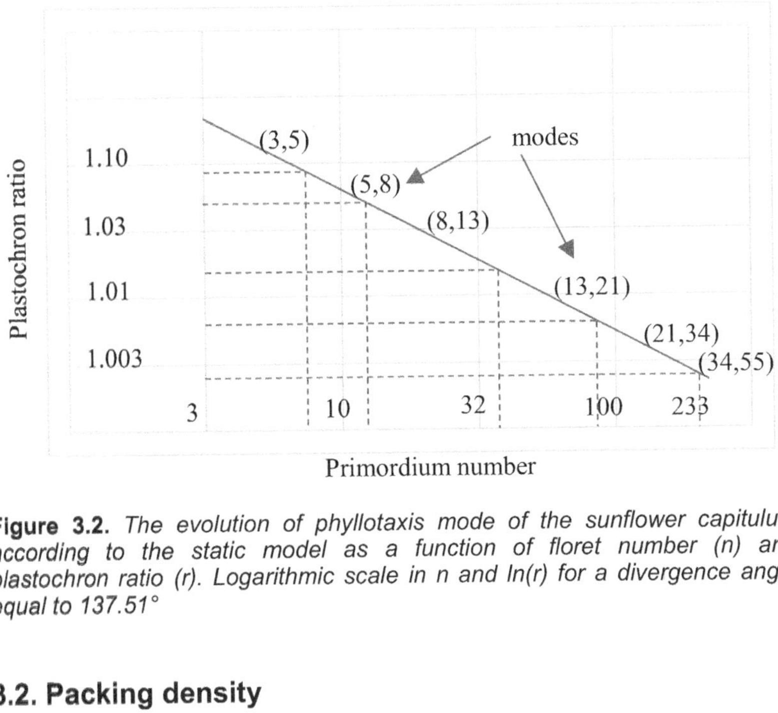

Figure 3.2. *The evolution of phyllotaxis mode of the sunflower capitulum according to the static model as a function of floret number (n) and plastochron ratio (r). Logarithmic scale in n and ln(r) for a divergence angle equal to 137.51°*

3.2. Packing density

For each organ, we define a "theoretical area" (A) that is proportional to the square of the distance to the nearest organ. The total theoretical area of the capitulum is the sum of the areas for each organ. These hypotheses allow us to compare the packing density of phyllotaxis with different divergence angles (Figure 3.3).

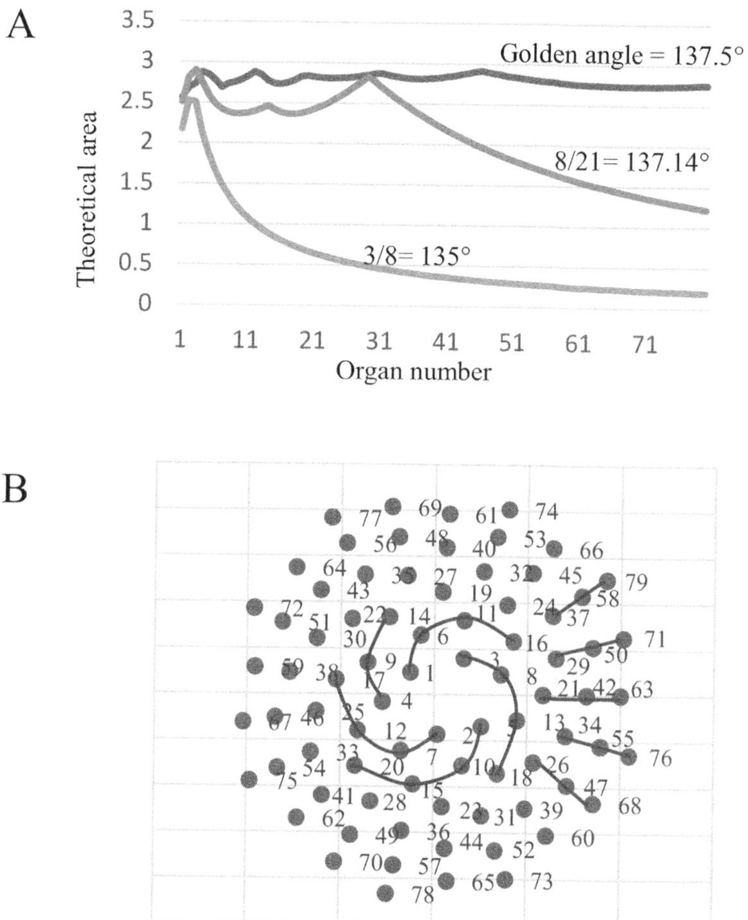

Figure 3.3. *(A) Theoretical area versus organ number with different divergence angles. (B) Organ arrangement for divergence angle = 8/21 x 360°*

When the divergence angle is a (rational) fraction p/q of a full circle (360°), for instance 8/21, starting from a given primordium (e.g. $n = 21$), the organ number n + q (here 42) is on the same radius as the organ we started with, after p turns (here 8) (Figure 3.3(B)):

$$\frac{p}{q} = \frac{\text{Number of turns}}{\text{Number of florets}} \qquad [3.3]$$

Thus, the phyllotaxis which was spiral for the first organs (1, 6, 11, 16) become whorled and in a centric representation forms a star whose branches represent the orthostichies (Figure 3.3(B)).

From organ q onwards (here 21), the angular difference between the neighboring organs is canceled out, which reduces the total distance between these two organs, and thus the theoretical area of the floret (Figure 3.3(A)). When phyllotaxis becomes whorled, the theoretical area of primordia collapses. The available space is thus poorly used by a whorled structure that is made up of numerous units. By contrast, in the case of golden (irrational) divergence angle, phyllotaxis remains spiral and the theoretical area keep high values.

In order to compare our results with those of Strauss et al. (2020), let us take for example a structure of medium complexity containing 30 organs whose size increases linearly as we move away from the apex (arithmetic growth or "plastochron difference").

By calculating the total theoretical area, we observe that certain particular values of divergence angles maximize packing efficiency, whereas others minimize it (local maxima and minimums) (Figure 3.4). The latter corresponds to rational divergence angle such as $1/3 \times 360° = 120°$, for which whorled arrangement is established from the onset.

By contrast, divergence angles that maximize packing efficiency are irrational. The divergence angle corresponding to the Fibonacci sequence (approximately 137.5°) produces a local maximum, however, other angles, such as those corresponding to the Lucas sequence (approximately 99.5°) and to the "first anomalous" are just as effective (Figure 3.4). These "noble angles" (Jean 1994) are listed by Marzec and Kappraff (1983).

Ridley (1982) defined packing efficiency as the ratio:

$$\frac{\text{Square of the distance between nearest neighbors}}{\text{Average area per organ}}$$

that for noble angles was equal to 82%.

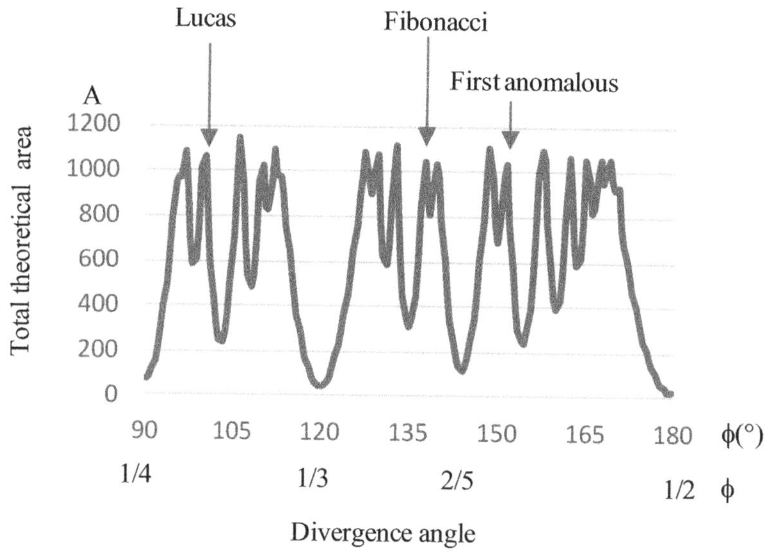

Figure 3.4. *The total theoretical area (A) as a function of the divergence angle (ϕ) for a structure containing 30 organs whose size increases linearly: unit is arbitrary*

3.3. Optimization of light capture

Strauss et al. (2020) aimed to determine whether natural selection could have favored spiral phyllotaxis produced by a divergence angle equal to the golden angle. They developed a computer model to calculate the light capture area (LCA) for different divergence angles in 3D plants bearing 2–30 leaves that resemble those of *Arabidopsis thaliana* (Brassicaceae) with different light incidence angles, leaf inclination angles and internode lengths.

LCA is the ratio of the leaf surface exposed to light over the total leaf surface: the less leaves shade other leaves, the greater this area is. In this model, the size of flowers increases linearly (Figure 3.5).

Strauss et al. (2020) showed that the golden angle (137.5°) optimizes light capture but other angles, such as the Lucas angle

(99.5°) do so as well, which would go against the idea that the golden angle is favored by natural selection. They noted the similarity between the effects of the divergence angle on light capture and packing efficiency in a regular lattice, as calculated by Marzec and Kappraff (1983).

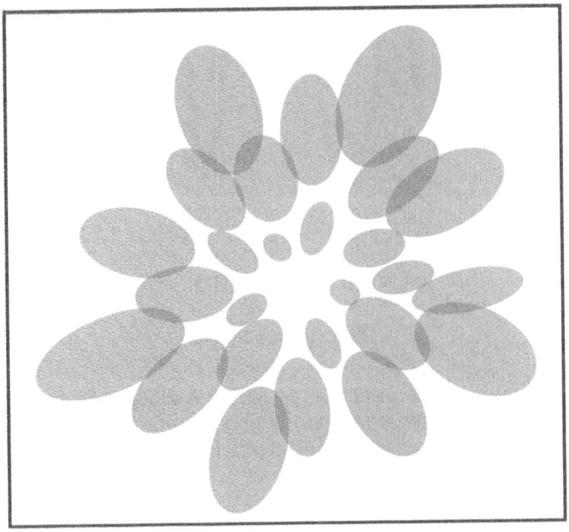

Figure 3.5. *Simulated plant with idealized leaves by Strauss et al. (2020). The leaves increase linearly in size*

The area units being arbitrary, we can establish a simple empirical correlation between the LCA of a fully grown plant, as modeled by Strauss et al. (2020), and the theoretical area of the meristem as calculated by our static model (Figure 3.6):

$$LCA = \ln(A) \qquad [3.4]$$

This relation has a correlation coefficient that is remarkably close to 99%. Thus, the light capture area behaves exactly like a bidimensional structure packed with disks that have constant divergence angles.

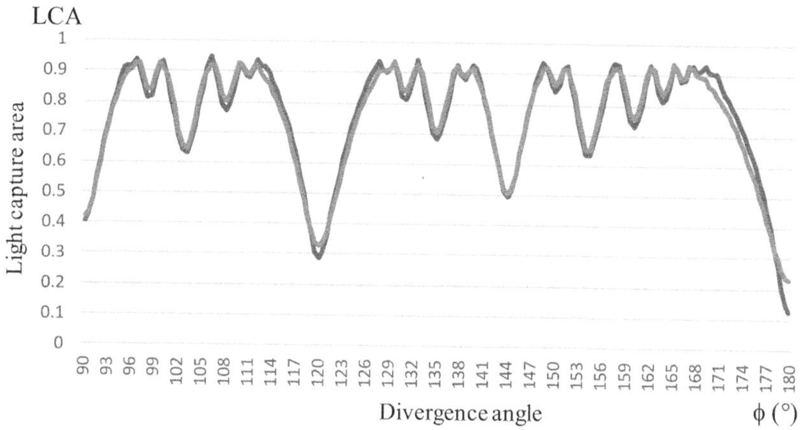

Figure 3.6. *The light capture area (LCA) for different divergence angles was calculated by Strauss et al.*[1] *(in red) and by applying the logarithm of the total theoretical area of the leaf meristem using our static model (in blue)*

3.4. Discussion

The analysis of self-shading caused by leaves on a plant shows a similarity with the final arrangement of organs in a bidimensional structure with constant divergence angles, such as the capitulum of a sunflower. In the case of light capture (with an angle of incident light of 15° and a leaf angle of 75°), the closer the nodes are, as in plants with fast growth, the more the phyllotaxis produced by irrational divergence angles are useful.

However, the Fibonacci sequence is not the only sequence that has these optimizing properties. It is also the case for sequences converging toward other "noble angles", the main one being the Lucas sequence. The golden angle does not ensure a global maximization of light capture but is only one angle among others that produce local maxima. We would like to find out why, with equivalent selective capabilities, non-Fibonacci phyllotaxis are less successful than those that converge toward the golden angle.

1 We thank Richard Smith for giving us access to his data.

Strauss et al. (2020) concluded that the iterative mechanism for organ positioning itself (what we call the phyllotaxic engine) is a more likely target for evolutionary pressure, rather than a specific divergence angle.

3.5. Appendix: the golden angle

Greek mathematicians studied what we today call irrational numbers, that is, numbers that cannot be expressed as the ratio of two integers (p/q). Around 300 BCE, Euclid defined what we today call the "golden ratio", as a geometrical proportion:

> A straight line is said to have been cut in extreme and mean ratio when, as the whole line is to the greater segment, so is the greater to the lesser.

The golden number is therefore the ratio (a/b) of the longer side (a) over the smaller side (b) of a rectangle such that it is also equal to the ratio of the sum of both sides (a + b) over the longer side (a). We can easily demonstrate (see Box 3.2) that the golden ratio is the solution to the following equation:

$$x^2 - x - 1 = 0 \qquad [3.5]$$

and equals:

$$x = \frac{1+\sqrt{5}}{2} \approx 1.6180 \qquad [3.6]$$

The expression of the golden ratio contains √5, which is an irrational number.

At the end of the 15th century, the Italian Franciscan friar Luca Pacioli referred to the golden ratio as a "divine proportion". During the 19th century, the golden ratio was considered by some as the most perfect expression of the architecture of human but also "natural works". However, contrary to popular belief, this divine proportion was seldom used by artists, architects or painters before the 19th century.

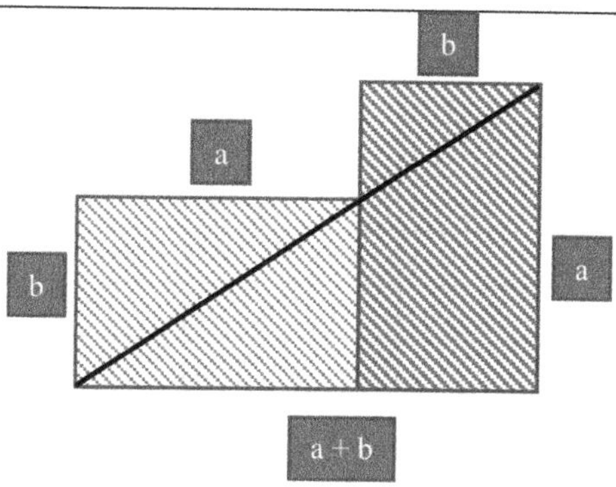

$$\frac{a}{b} = 1 + \frac{b}{a}$$

$$x = 1 + \frac{1}{x}$$

$$x^2 = x + 1$$

$$x^2 - x - 1 = 0$$

$$x = \frac{1 + \sqrt{5}}{2} \approx 1{,}6180$$

Definition of the golden ration
by Euclid (300 BCE):

$$x = \frac{a}{b} = \frac{a+b}{a}$$

Box 3.2. *The golden rectangle*

In the 13th century, the Italian mathematician Leonardo Fibonacci of Pisa defined a sequence of numbers that bears his name: the first two terms (F1 and F2) are equal to 1 and 2, respectively (initiating the sequence). The recurrence relation that defines the next terms is the following: each term is the sum of its two predecessors. Thus, the third term is equal to 1 + 2 = 3, the fourth 3 + 2 = 5, etc.

$$F_{n+2} = F_{n+1} + F_n \qquad [3.7]$$

Mathematicians from the Renaissance showed that the ratio of two consecutive terms in the Fibonacci sequence converge toward the golden ratio:

$$\frac{lim}{n \to \infty} \frac{F_{n+1}}{F_n} = \frac{1+\sqrt{5}}{2} \approx 1.6180 \qquad [3.8]$$

n	Fn	Fn + 1	Fn + 1/Fn	Fn + 1/Fn
1	1	2	2/1	2
2	2	3	3/2	1.5
3	3	5	5/3	1.666666667
4	5	8	8/5	1.6
5	8	13	13/8	1.625
6	13	21	21/13	1.615384615
7	21	34	34/21	1.619047619
8	34	55	55/34	1.617647059
9	55	89	89/55	1.618181818
10	89	144	144/89	1.617977528
11	144	233	233/144	1.618055556
12	233	377	377/233	1.618025751
13	377	610	610/377	1.618037135
14	610	987	987/610	1.618032787

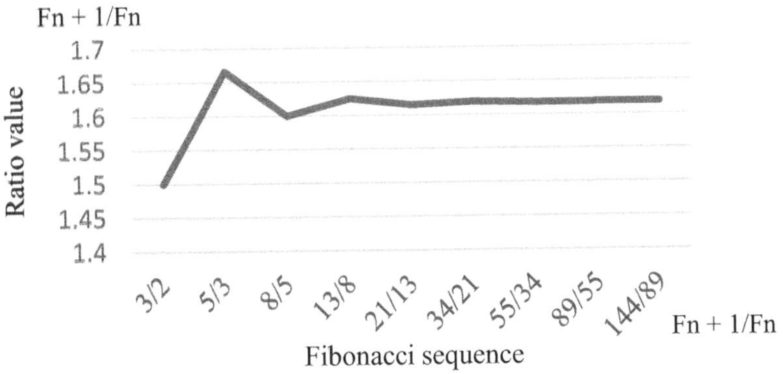

Figure 3.7. *Convergence of the Fibonacci sequence toward the golden ratio*

The golden angle (b) divides a circle (360°) in two parts, a and b, such that the ratio a/b is equal to the golden ratio ~ 1.62, so $\phi \sim 137.51°$.

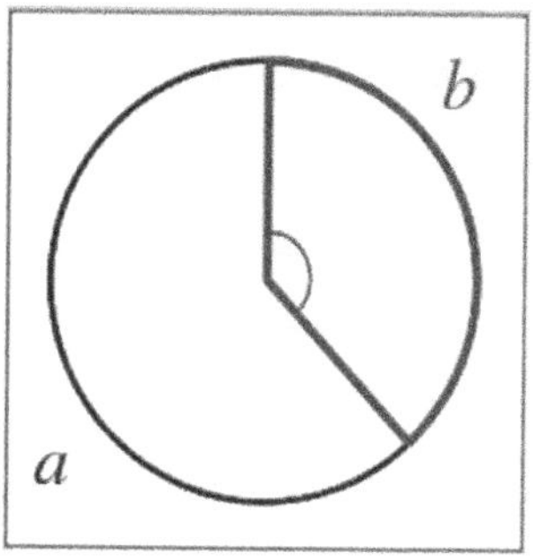

$$\frac{a}{b} = \frac{1+\sqrt{5}}{2} \quad \text{and} \quad a + b = 1$$

$$\varphi = b \, x \, 360° = \frac{3-\sqrt{5}}{2} \, x \, 360° \sim 137.51°$$

Box 3.3. *The golden angle*

Bravais and Bravais (1837a) carefully examined the fibers along stems being stripped of its bark and used a compass to measure the distance between these fibers. They showed that the correspondence between the leaves after p turns is not exact. In brooms and jasmin, leaf 6 deviates from leaf 1 by approximately 30°. In *Osyris*, leaf 8 deviates by more than approximately 20°.

For the Frenchmen, the divergence angle is constant in all plants and is comprised between 137° and 138°. More precisely, the angle would be equal to the limit of the ratio F_n/F_{n+2}, that is:

$$\frac{lim}{n\rightarrow\infty}\frac{F_n}{F_{n+2}} = \frac{lim}{n\rightarrow\infty}\frac{F_n}{F_{n+1}}\frac{F_{n+1}}{F_{n+2}} = \frac{3-\sqrt{5}}{2} \approx 0.382 \qquad [3.9]$$

which expressed in degrees is 137°38′28″. This angle is the one that divides a circle (360°) in "extreme and mean ratio", which is to say by the golden ratio.

Some plants are arranged according to sequences that are different from the Fibonacci sequence, such as 1, 3, 4, 7, 11, etc. (the Lucas sequence) in *Sedum reflexum*, whose divergence angle is 99°39′, or *Dipsacus sylvestris* which has 18, 29 and 47 parastichies. Its corresponding angle converges toward an irrational number that is equal to:

$$\frac{5-\sqrt{5}}{10} \qquad [3.10]$$

The sequence 2, 5, 7, 12, 19, etc. (the so-called first anomalous sequence) can be observed on the floral spike of *Bromelia ananas*, the cone of the maritime pine, *Dipsacus sylvestris* and *Dipsacus fullonum*. The corresponding angle converges toward:

$$\frac{7+\sqrt{5}}{22} \qquad [3.11]$$

that is, 151°8′8″.

Let us calculate the following continuous fraction, starting from the bottom.

$$x = 1 + \cfrac{1}{1 + \cfrac{1}{1 + \cfrac{1}{1 + \cfrac{1}{1 + \frac{1}{1}}}}}$$

(from bottom to top: → 2, → 1.5, → 1.6667, → 1.6, → 1.625)

The number under the lowermost fraction bar is equal to $u_1 = 1 + 1/1 = 2$. The number under the fraction bar above it is equal to: $u_2 = 1 + 1/2 = 1.5$. The number under the fraction bar above is equal to: $u_3 = 1 + 1/1.5 = 1.6667$. The number under the last fraction bar is equal to: $u_4 = 1 + 1/1.6667 = 1.6$. Let us calculate one more term, namely $u_5 = 1 + (1/1.6) = 1.625$.

By increasing the number of iterations, we climb the Fibonacci sequence and get a better approximation of the golden ratio.

Box 3.4. *Continuous fractions (Bravais and Bravais 1837a)*

4

The Dynamical Model

4.1. Description of the model

Our dynamical model is based on the standard model of Douady and Couder (1996a) and describes an inhibition potential that controls primordia initiation at the meristem *front* (Figures 1.3(B) and 1.6). Plant organs such as leaves, sepals, petals, stamens and carpels originate from the shoot apical meristem (SAM) as small primordia that arise *at the periphery of the central zone (CZ) of the meristem* (Godin et al. 2020) at a given distance from the tip (apex) of the meristem before migrating away from the CZ (Figures 1.3 and 4.1).

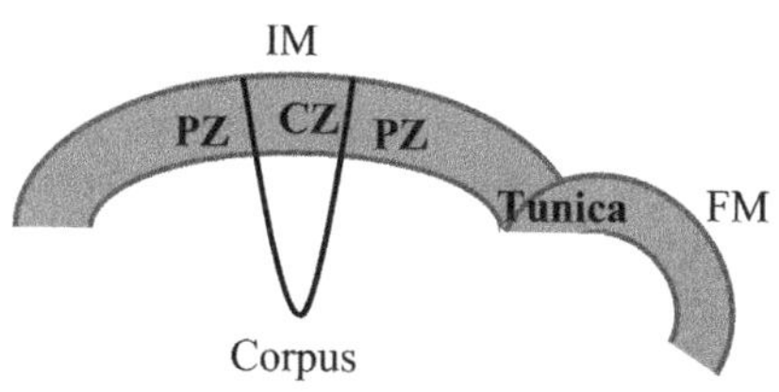

Figure 4.1. *Schematic representation of an inflorescence meristem (IM) and a floral meristem (FM). The IM can be divided into two distinctive zones: the central zone (CZ), which contains stem cells, and the peripheral zone (PZ) flanking the CZ, where FMs form. The upper layer of the tunica (the outer layer of the meristem) forms the epidermis, under which the medullary parenchyma is formed (redrawn from Alvarez-Buylla 2010)*

For a color version of all the figures in this book, see www.iste.co.uk/walch/phyllotaxis.zip.

According to the *biochemical interpretation* of this model, accumulation of auxin (a phytohormone) induces organ initiation at the CZ rim. Auxin is mainly synthetized in the young organs and is actively transported at the meristem tip through membrane transporters of the PIN1 family. Local accumulation of auxin which is identified with the abstract *inhibitory field* hypothesized in the model triggers organ initiation, subsequently depleting auxin around organs: a meristem tip is abundant with auxin, except *around* the places where young organs have been initiated (Galván-Ampudia et al. 2020).

The front of the meristem can be represented on a plane by a circle of fixed radius R_0, chosen as unit of length. Primordia are represented as points located between this circle and the outer edge of the meristem. A new primordium (i) initiates on this circle at a point of polar coordinates (R_0, θ), where θ is the angle that minimizes the *inhibition potential* U_i generated by preexisting primordia (Figure 4.2(A)).

Consider a meteoroid located 400 km away from the Earth: the force of universal attraction pulls the meteoroid toward the center of our planet. In the absence of this meteoroid, something remains: the attraction field created by the Earth. The gravitational potential is a measure of this attraction *field* and it takes a second object, like a meteoroid, for the potential force to materialize into an actual force.

Like universal attraction, the inhibition generated by a primordium decreases as the distance from that primordium increases. The decay length is a parameter that determines the rate of this decrease (Figure 4.2(B)). Maps of inhibition potentials show the intensity of the inhibition around each source (Figure 4.2(A)): the intensity is high near the center of each primordium and becomes weaker as the distance from each center increases. In Figure 4.2(A), a new primordium (number 30 shown by a star) will be initiated on the front between primordia 22 and 25, where the inhibition potential is the lowest.

Box 4.1. *What is a potential?*

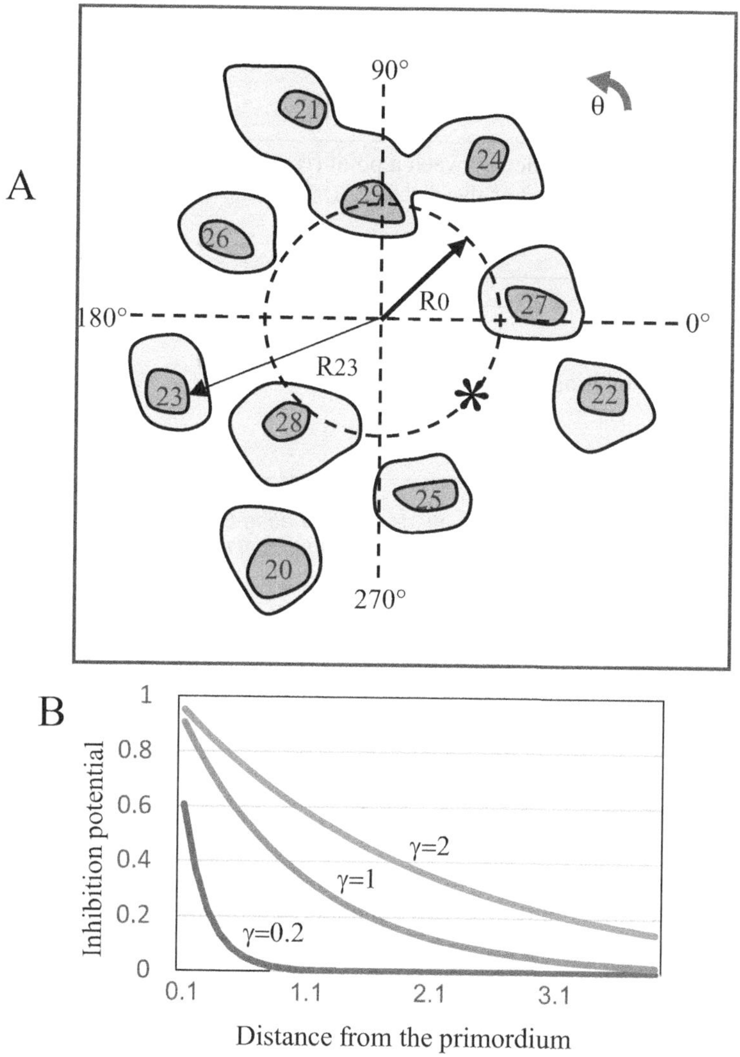

Figure 4.2. *(A) Map of the inhibition field on the vegetative meristem of Linum usitatissimum on the 15th day of bud development at the initiation of the 30th primordium (located between primordia 22 and 25) according to the dynamical model. The location of each primordium (i) is specified by its distance from the center of the meristem (R_i) and its angular coordinate (θ_i). The radius of the front where primordia arise (R_0) is taken as unit of length. (B) Variation in the range of the inhibition field for three different decay lengths*

The inhibition potential decreases exponentially with the distance from the primordium generating the inhibition field, the decay length being γ (equations [4.1] and [4.2]).

If d_{ij} is the distance between a point (i) with coordinates (R_i, θ_i) and n primordia (j) located at the distance R_j from the center of the meristem with an angular coordinate θ_j:

$$d_{ij} = \sqrt{R_i^2 + R_j^2 - 2R_iR_j\cos(\theta_i - \theta_j)} \qquad [4.1]$$

then the inhibition potential is given by:

$$U_i(R_i, \theta_i) = \sum_{j=1}^{n} \exp(-\frac{d_{ij}}{\gamma}) \qquad [4.2]$$

Box 4.2. *Equations determinating the location of primordia in the dynamical model*

Our model differs from those of Douady and Couder (1996b), Smith et al. (2006) and Kitazawa and Fujimoto (2015), which assume that a constant threshold of potential determines primordium initiation and plastochron ratio. Here, the parameter imposed is not the threshold potential value, but the plastochron ratio between all primordia.

From the positions of existing primordia, it is possible from equations ([4.1] and [4.2]) to calculate the inhibition potential at any point on the meristem, and in particular on the front knowing that at initiation the radial position is $R_0 = 1$. The angular position of a new primordium (i) is *where the inhibition potential on the front is the lowest*. After the initiation of a new primordium, all existing primordia move away from the apex by a distance that is determined by the plastochron ratio r_i (equation [1.1]). Taking into account these new positions, the inhibition potential at any point of the meristem can be calculated de novo to determine the position of primordium (i+1), etc.

This iterative process must be initialized: at the start, the positions of a few primordia must be given as input for the calculations. For example, for a vegetative meristem, the position of the two cotyledons, or for a reproductive meristem, that of bract and bracteoles.

Upon request, the authors can provide an Excel spreadsheet where the initial position of two to four primordia can be set, and the positions of the following six primordia can be calculated (Eichler.xls). To request the spreadsheet please contact Jean-Paul Walch at jpwalch511@gmail.com.

In addition to initial setting of the coordinates of first primordia (R_k, θ_k), the model requires plastochron ratio (r_i) data at the initiation of all the calculated primordia.

Box 4.3. *Free computer software*

Scaling the parameters by the constant c:

$x \rightarrow cx$; $y \rightarrow cy$ and $\gamma \rightarrow c\gamma$ that is $d_{ij} \rightarrow cd_{ij}$ and $\gamma \rightarrow c\gamma$

does not change the inhibition potential, that is, the model is scale invariant:

$$U(\theta) \rightarrow U(\theta)$$

The dimensionless quantity characterizing the range of the inhibition interaction is the "effective distance":

$$\delta = \frac{R_0}{\gamma}$$

which is the ratio of the actual distances divided by the decay length of the interaction. The model thus produces the same pattern in plants with the same effective distance, irrespective of the real diameter of their meristem.

Box 4.4. *Effective distance*

We modeled with the dynamical model the vegetative meristem of *Linum usitatissimum* using the plastochron ratio data obtained by Williams (1975; Table 1.1) with R_0, the radius of the front, the same for all iterations (Walch 2022). This model minimizes the inhibition potential at the front at each iteration, from which the angular position of each new primordium is determined.

All calculated points representing primordia are located inside their respective observed primordia (Figures 1.3 and 1.6); there is no cumulative error and even the error gets reduced with the primordium number, showing the good fit of our model to the observations of the vegetative meristem. In addition, the divergence angle rapidly converged toward the golden angle (Figure 4.3).

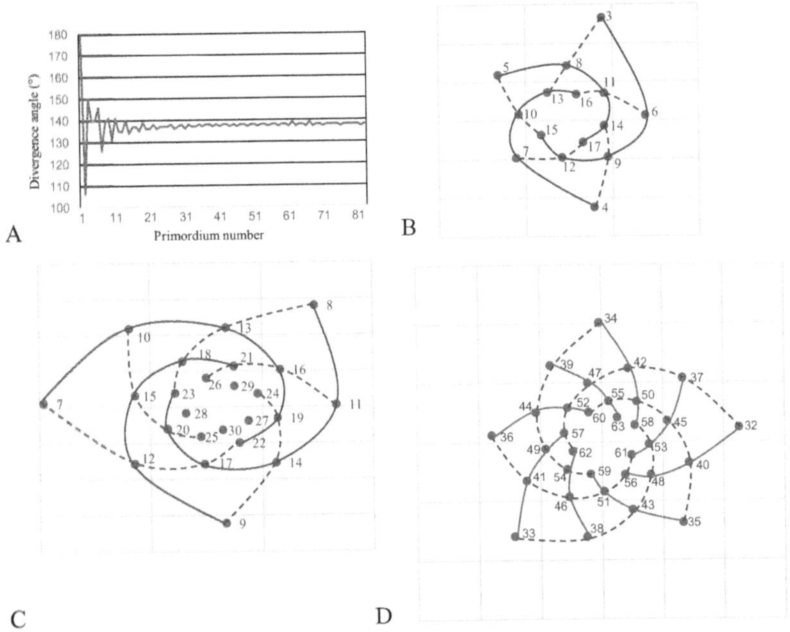

Figure 4.3. *(A) Value of the divergence angle from model between successive primordia for 81 primordia on the vegetative meristem of Linum usitatissimum. (B--D) Model of the vegetative meristem of L. usitatissimum at different developmental stages: solid and dashed lines are anti-clockwise and clockwise parastichies, respectively; plastochron ratios from Table 1.1 were used for the parametrization of the model. Phyllotaxis modes are respectively (3, 5) in B (primordia 3--17; r = 1.08) and C (primordia 7--30; r = 1.06) and (5, 8) in D (primordia 32--63; r = 1.05).* $R_0 = 1$; $\gamma = 0.2$

4.2. Phyllotaxis modes (analytical model)

In the case of Fibonacci phyllotaxis, the divergence angle between primordia is approximately constant and equal to 137.5° after about 20 primordia (Figure 4.3(A)). It is then possible to calculate the inhibition potential *analytically* starting from an initial primordium i = 0 ($R_0 = 1$, $\theta_0 = 0$) (Table 4.1). The angular position of existing primordia (j) is a multiple of the golden angle (ranging from 0° to 360°). The radial position R_j is equal to r (the plastochron ratio) to the power of j (the primordium number): if r is constant, at each initiation the distance of all the existing primordia is multiplied by r. The distance d_{ij} is calculated by equation [4.1] and the inhibition potential produced by primordium j by the formula of equation [4.2]. So, the total inhibition $U_i(R_0, \theta_0)$ is the sum of all U_{ij}.

j	Angle (θ_j)	$\cos(\theta_j)$	R_j	Distance d_{ij}	Inhibition U_{ij}
1	137.5	–0.737	1.040	1.90	0.00007
2	275	0.087	1.082	1.41	0.00088
3	52.5	0.609	1.125	0.95	0.00881
4	190	–0.985	1.170	2.16	0.00002
5	327.5	0.843	1.217	0.65	0.03796
6	105	–0.259	1.265	1.80	0.00012
7	242.5	–0.462	1.316	1.99	0.00005
8	20	0.940	1.369	0.55	0.06439
9	157.5	–0.924	1.423	2.38	0.00001
10	295	0.423	1.480	1.39	0.00095
11	72.5	0.301	1.539	1.56	0.00040
12	210	–0.866	1.601	2.52	0.00000
13	347.5	0.976	1.665	0.72	0.02705
14	125	–0.574	1.732	2.45	0.00000
15	262.5	–0.131	1.801	2.17	0.00002
16	40	0.766	1.873	1.28	0.00166

j	Angle (θ_j)	$\cos(\theta_j)$	R_j	Distance d_{ij}	Inhibition U_{ij}
17	177.5	–0.999	1.948	2.95	0.00000
18	315	0.707	2.026	1.50	0.00056
19	92.5	–0.044	2.107	2.37	0.00001
20	230	–0.643	2.191	2.94	0.00000
21	7.5	0.991	2.279	1.29	0.00155
Total Ui					0.14452

$\phi = 137.5°$; $\gamma = 0.2$; $r = 1.04$; $R_0 = 1$.

Table 4.1. *Analytical calculation of inhibition potential from equations [4.1] and [4.2] (Fibonacci angle; r = 1.04)*

> The Eichler.xls program also allows the calculation of the polar coordinates and the inhibition potential of each primordium for all the values of γ, ϕ, r and R_0 using the analytical method (as in Table 4.1).

Box 4.5. *Free computer software*

The data in Table 4.1 are sufficient to explain the existence of parastichies (8, 5) when the plastochron ratio equals 1.04, since the distance between primordium 0 and primordium 8 is 0.55, and 0.65 between primordium 0 and primordium 5, all other primordia being further away from primordium 0.

Starting from any primordium (0), as we move away on the generative spiral (i.e. j increases), the radial distance Rj increases. However, primordium (1), which is the closest radially to primordium (0), is 137.5° angularly distant from it. Primordium (3) is angularly closer to primordium (0), since its angular distance from primordium (0) is $3 \times 137.5° - 360° = 52.5°$ (one full turn +52.5°) but it is radially further from primordium (0) than primordium (1). The distance d_{ij} is therefore a compromise between radial distance and angular distance. The radial distances depend on the plastochron ratio: this is the reason why the number of parastichies (m, n) depends on the plastochron ratio.

For r = 1.04, primordium 5 is at the angular distance 327.5° = –32.5° (–32.5° from one full turn). It is therefore quite close in degrees to primordium (0); primordium 8 is even closer, its angular distance being 20°, which explains the proximity of these two primordia and therefore the (5, 8) phyllotaxis mode.

If r = 1.03, the radial distances R_j are lower than for r = 1.04 (Table 4.2). Smaller angular distances (θ_{13} = 347.5° – 360° = –12.5°) determine one of the minimum total distances. The phyllotaxis mode thus becomes (8, 13), primordium 13 having a smaller angular distance to primordium 0 than primordium 5.

j	Angle (θ_j)	$\cos(\theta_j)$	R_j	Distance d_{ij}	Inhibition U_{ij}
1	137.5	–0.737	1.030	1.89	0.00008
2	275	0.087	1.061	1.39	0.00094
3	52.5	0.609	1.093	0.93	0.00959
4	190	–0.985	1.126	2.12	0.00003
5	327.5	0.843	1.159	0.62	0.04432
6	105	–0.259	1.194	1.74	0.00016
7	242.5	–0.462	1.230	1.91	0.00007
8	20	0.940	1.267	0.47	0.09384
9	157.5	–0.924	1.305	2.26	0.00001
10	295	0.423	1.344	1.29	0.00156
11	72.5	0.301	1.384	1.44	0.00073
12	210	–0.866	1.426	2.35	0.00001
13	347.5	0.976	1.469	0.54	0.06798
14	125	–0.574	1.513	2.24	0.00001
15	262.5	–0.131	1.558	1.96	0.00006
16	40	0.766	1.605	1.06	0.00508
17	177.5	–0.999	1.653	2.65	0.00000
18	315	0.707	1.702	1.22	0.00223
19	92.5	–0.044	1.754	2.06	0.00003

j	Angle (θ_j)	$\cos(\theta_j)$	R_j	Distance d_{ij}	Inhibition U_{ij}
20	230	–0.643	1.806	2.57	0.00000
21	7.5	0.991	1.860	0.88	0.01236
Total Ui					0.14452

$\phi = 137.5°$; $\gamma = 0.2$; $r = 1.03$; $R_0 = 1$.

Table 4.2. *Analytical calculation of inhibition potential from equations [4.1] and [4.2] (Fibonacci angle; r = 1.03)*

These analytical calculations can be carried out for all values of divergence angles and plastochron ratios: we then obtain a map of the phyllotaxis modes that is equivalent to van Iterson's diagram (Figure 2.2(B)), and which we call "the phyllotaxis frieze" (Figure 4.4).

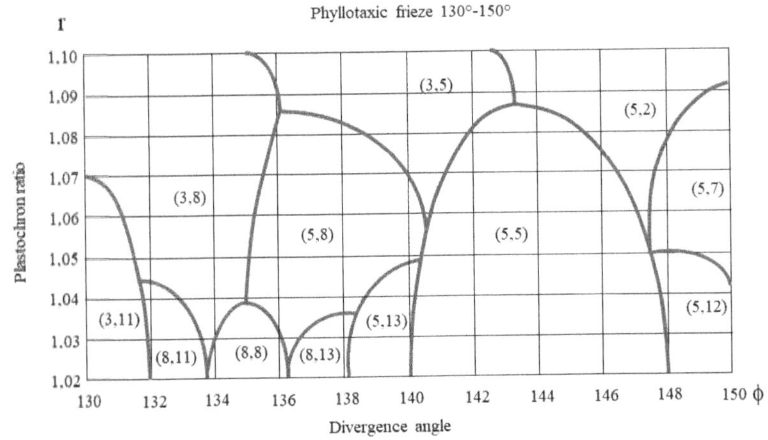

Figure 4.4. *The phyllotaxis frieze delimiting the different regions corresponding to different phyllotaxis modes for divergence angles ranging between 130° and 150°*

The *phyllotaxis frieze* is easier to read than van Iterson's diagram, which uses a growth index as ordinate and offers a practical application: it allows us to determine a phyllotaxis mode from the pair (ϕ, r) and vice versa a region of (ϕ, r) coordinates from a phyllotaxis mode. In addition, it highlights the regions where whorled phyllotaxis, such as (5, 5) or (8, 8), ensures a dense distribution of primordia.

van Iterson's diagram is suitable for mathematicians; the phyllotaxis frieze provides information that is valuable to botanists, the plastochron ratio being a directly measurable quantity, which is not the case for the growth index G.

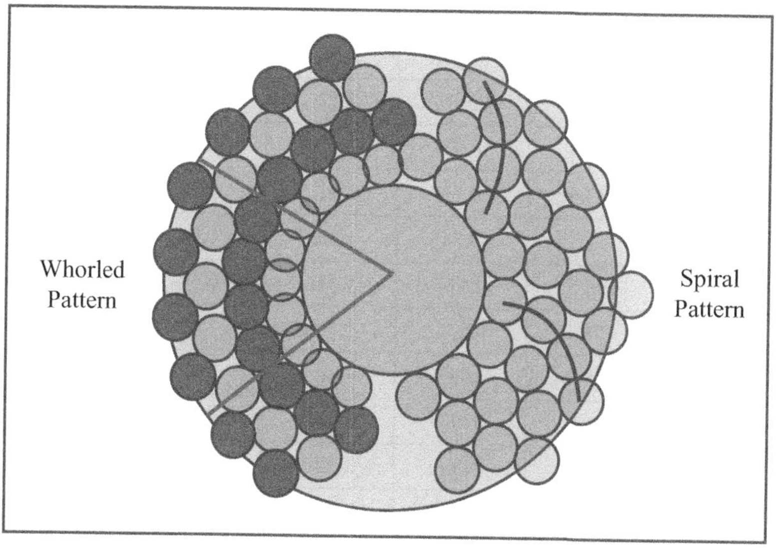

Figure 4.5. *Packing of disks of equal diameter within a ring (from Green 1999). In a whorled arrangement, disks close to the center are forced to overlap and disks far from the center cannot join each other. On the contrary, packing is regular in a Fibonacci spiral pattern*

Douady and Couder (1996c) reduced the growth index (the plastochron ratio) starting with a value corresponding to the decussate mode (2, 2). In most cases, the model produced Fibonacci phyllotaxis ((2, 3) mode). However, very rarely, and for specific values of the parameters, they obtained bijugate phyllotaxis (2, 4), Lucas phyllotaxis (3, 4) and even the second accessory pattern (4, 5).

Atela (2011) found that in a disk stacking model, Fibonacci, Lucas, bijugate and multijugate patterns are all variations of the same unifying phenomenon and that the difference is only due to small changes in the position of the initial primordia.

Box 4.6. *Transition to the Lucas series (Douady and Couder 1996c; Atela 2011)*

For a divergence angle equal to the Lucas angle (99.5°) and a plastochron ratio of 1.04, the phyllotaxis mode is (7, 11) (Table 4.3).

i-j	Angle	cos(θ-θj)	R_j	Distance d_{ij}	Inhibition U_{ij}
1	99.5	–0.165	1.040	1.56	0.00042
2	199	–0.946	1.082	2.05	0.00003
3	298.5	0.477	1.125	1.09	0.00426
4	38	0.788	1.170	0.72	0.02672
5	137.5	–0.737	1.217	2.07	0.00003
6	237	–0.545	1.265	1.99	0.00005
7	336.5	0.917	1.316	0.56	0.05961
8	76	0.242	1.369	1.49	0.00059
9	175.5	–0.997	1.423	2.42	0.00001
10	275	0.087	1.480	1.71	0.00019
11	14.5	0.968	1.539	0.62	0.04421
12	114	–0.407	1.601	2.21	0.00002
13	213.5	–0.834	1.665	2.56	0.00000
14	313	0.682	1.732	1.28	0.00167
15	52.5	0.609	1.801	1.43	0.00078
16	152	–0.883	1.873	2.80	0.00000
17	251.5	–0.317	1.948	2.46	0.00000
18	351	0.988	2.026	1.05	0.00525
19	90.5	–0.009	2.107	2.34	0.00001
20	190	–0.985	2.191	3.18	0.00000
21	289.5	0.334	2.279	2.16	0.00002
Total Ui					0.14386

$\phi = 99.5°$; $\gamma = 0.2$; $r = 1.04$; $R_0 = 1$.

Table 4.3. *Analytical calculation of inhibition potential from equations [4.1] and [4.2] (Lucas angle; r = 1.04)*

5

Molecular or Contact Pressure Origin?

5.1. The dynamics of phyllotaxis

We calculated the values of the inhibition potential from the divergence angle (ϕ) and the plastochron ratio (r) using the analytical model (each value is calculated from equations [4.1] and [4.2] with ϕ = const. and r = const.).

> Our inhibition potential map (Figure 5.1) is similar to the gravitational potential maps drawn by astrophysicists. At any point on their map, the gravitational force will lead massive bodies down the steepest slope (the gradient) of this potential.
>
> In the phyllotaxis model, the trajectories, under the effect of the inhibition field generated by primordia, climb the valleys of the inhibition potential as the plastochron ratio decreases.

Box 5.1. *Force derived from a potential*

Primordia initiate along the valleys (local minimums) of the inhibition potential. The dynamics of the system as the plastochron ratio decreases is consistent with that of van Iterson's diagram (Figure 2.2): the same bifurcations are found, the number of inhibition

For a color version of all the figures in this book, see www.iste.co.uk/walch/phyllotaxis.zip.

potential minimums increasing with a decrease in the value of the plastochron ratio (Figure 5.1). The system rises in the Fibonacci sequence by a succession of "quasi-bifurcations". Indeed, at each bifurcation, such as from (5, 8) to (8, 13), the passage to (13,5) is impossible, because these two types of parastichies have the same chirality (direction of rotation).

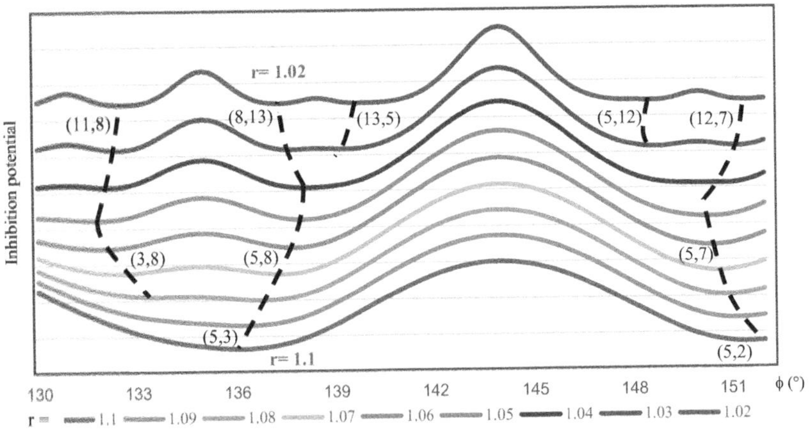

Figure 5.1. *The inhibition potential (ln (U)) as a function of the divergence angle (ϕ) for different plastochron ratios (r) calculated from the the analytical model. For each r value, the curve of the natural logarithm of the inhibition potential is shown; as r decreases, trajectories in phase space climb the valleys formed by the inhibition potential curves (dashed lines)*

Our model gives a physicochemical basis to van Iterson's diagram (Walch 2022). For example, the vegetative meristem of *Linum* goes from a (3, 5) to a (5, 8) mode between primordium 17 and primordium 63 when the plastochron ratio decreases from 1.08 to 1.045 (Figure 4.4).

5.2. The molecular origin of phyllotaxis

The study of the biochemical origin of phyllotaxis has been primarily focused on the inflorescence meristem (IM) of *Arabidopsis thaliana* (Brassicaceae).

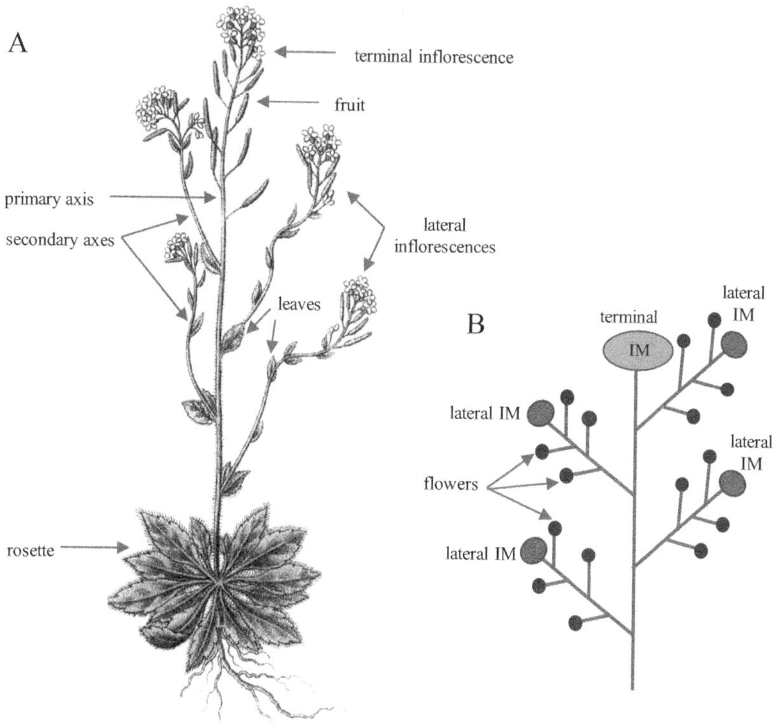

Figure 5.2. *Arabidopsis thaliana inflorescence architecture. (A) Drawing from Johann Georg Sturm (1796). (B) Diagram. IM = inflorescence meristem*

Arabidopsis thaliana inflorescence is a compound raceme (Figures 5.2 and 7.5). During the vegetative phase, the terminal meristem (IM) gives rise to rosette leaves. Subsequently, during the reproductive phase, the inflorescence forms: IM primordia initiate cauline leaves and then floral meristems (FM) where floral organs will initiate.

This initiation is controlled by the accumulation of the phytohormone auxin. Auxin accumulated in a new primordium becomes depleted in the vicinity of that primordium, thus inhibiting the formation of another primordium nearby (Galván-Ampudia et al. 2020). *A. thaliana* FM are initiated in the peripheral zone (PZ) of the inflorescence apical meristem (IM) with golden divergence angle and form Fibonacci spirals (Figure 5.3(A); Galvan-Ampudia et al. 2020).

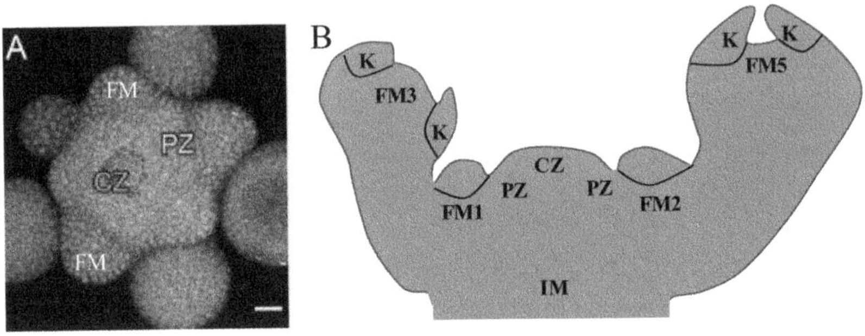

Figure 5.3. *The inflorescence meristem of Arabidopsis: (A) horizontal section (centric representation) showing the central zone (CZ) and the peripheral zone (PZ) of the inflorescence meristem (IM) and surrounding floral meristems (Galván-Ampudia et al. 2020). (B) Lateral section of the IM showing floral meristems (FM) from the youngest (number 1) to the oldest (number 5), (4 is hidden from view) (redrawn from Alvarez-Buylla (2010)). Oldest FM numbers 3 and 5 have already initiated sepals (K)*

From equations [4.1] and [4.2], the plastochron ratio definition [1.1] and Table 1.1 (plastochron ratios), we were able to calculate the location of primordia inside the IM (Figure 5.4(A)) and the inhibition potential at the front (Figure 5.4(C)) at the initiation of primordium number P0 = 30 (when the divergence angle is 137.5°, according to Figure 4.4; r = 1.07; mode is (5,8)). The local potential maxima correspond to the primordia already initiated and the minimums of this potential at the front are the places where future primordia will form.

Galvan-Ampudia et al. (2020) used the DII-VENUS biosensor to visualize the spatial distribution of auxin concentration within the IM of *Arabidopsis*. Our modeling of the inhibition potential on the front (Figure 5.4(C)) matched their data for the distribution of the concentration of the auxin reporter (Godin et al. 2020, their figure 8), except that the qDII maximum produced by the latest primordium (primordium P1) was lower than that of its predecessor (primordium P2), even though the latter is further away from the front, which should have reduced the potential generated by it. An explanation is that the dynamics of auxin create a delay in the buildup of the inhibition potential. To resolve this problem, it would be necessary to study the dynamics of this system on a timescale that is much shorter than a plastochron.

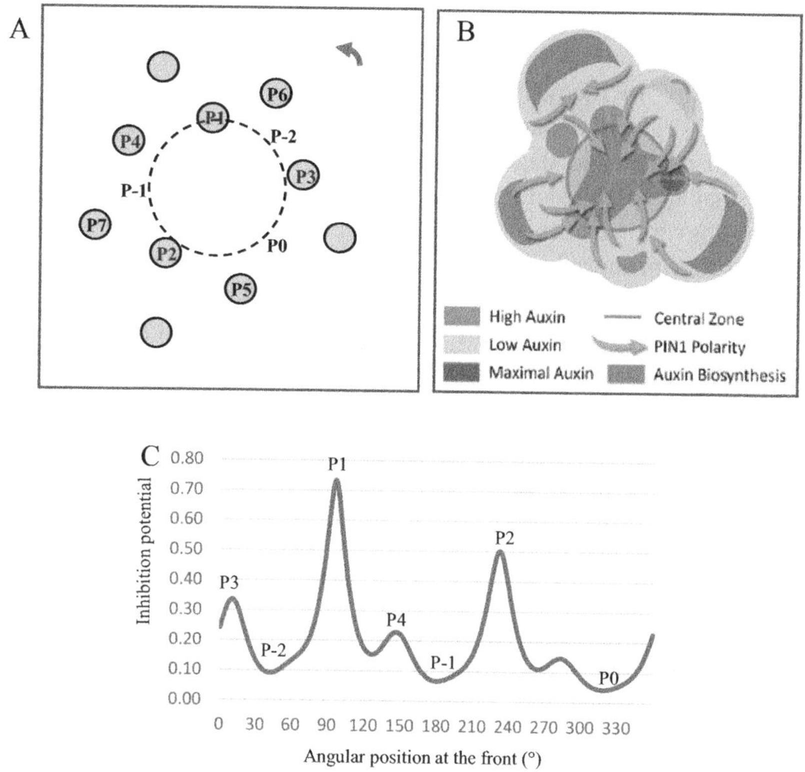

Figure 5.4. *(A) Location of floral meristems of A. thaliana according to the model. P1 is the position of the last primordium to have been initiated. Primordia Pn are its predecessors, and P0, P-n its successors. (B) Schematic representation of the tissue-scale organization of auxin transport and biosynthesis in relation to auxin distribution (Galván-Ampudia et al. 2020). (C) Profile of inhibition potential on the meristem front according to our model*

Galvan-Ampudia et al. (2020) showed that the dynamics of auxin do not result purely from cellular growth, the maxima surpassing the average displacement of individual nuclei. Indeed, auxin is transported across tissues. PIN1 proteins are auxin carriers whose polarity (the difference in intensity across membranes on both sides of the cell interface) determines the direction of auxin flux. The polarity of the cell interface determines the direction in which each cell interface locally contributes to orient the flow of auxin.

The vector map shows a strong convergence of PIN1 polarity toward the center of the SAM (Figure 5.4B from Galvan-Ampudia et al. 2020). In the vicinity of P7, there is an inversion in PIN1 polarity at organ boundaries, which isolates the flower from the rest of the SAM at this late stage. P3 to P5 show a general flux toward the SAM that is locally deflected around the zones of auxin minimums before converging back toward the center of the SAM. These results suggest that primordia P3-P5 are primarily auxin producers, and that P-1 and P0 act as attractors of this auxin flux. The CZ can be an area of auxin maxima as it is at the confluence of centripetal auxin fluxes. This confluence may divert auxin away from certain areas resulting in the appearance of auxin minimums.

The biochemical inhibition field therefore does not propagate at a great distance from the front, and the ability of the model to take into account the repulsion between organs located at these distances must be attributed to other physical forces (see Conclusion).

Like most Brassicaceae, *Arabidopsis thaliana* has flowers that are composed of four sepals, four petals, six stamens and two carpels. Flower development in *A. thaliana* has been divided into 12 stages by Smyth et al. (1990). During stage 1, bulges appear at the edge of the IM. At stage 2, the FM separate from the IM: their initiation is on the same Fibonacci spiral (with a divergence angle of 137.5°) as the leaves. Then, the sepal primordia are initiated (stage 3) and grow until they cover the floral meristem (stage 4). Petal and stamen primordia are initiated at stage 5 and are enclosed by the sepals (stage 6; Figure 5.5).

Remarkably, among angiosperms, *Arabidopsis*, like most Brassicaceae, lacks bracts. However, an ephemeral bract does appear at the start of development but quickly become sepal.

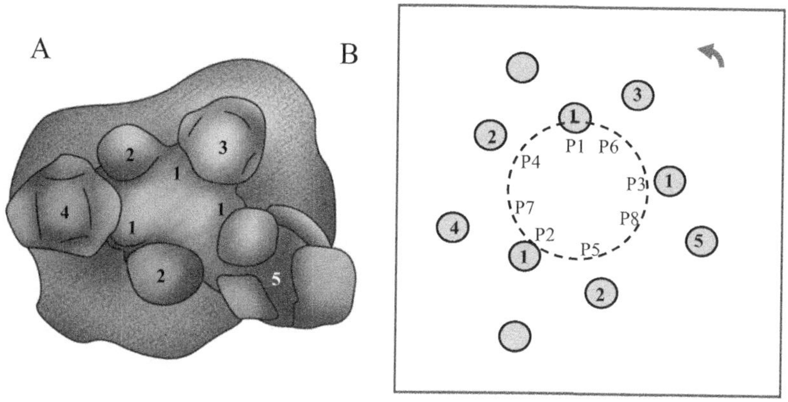

Figure 5.5. *(A) The inflorescence meristem of Arabidopsis thaliana redrawn from the SEM micrographs of Smyth et al. (1990). Numbers indicate the developmental stage of floral meristems. At stage 3, we begin to distinguish the tetramerous arrangement of the sepals; four sepals are clearly visible at stage 5. (B) Location of floral meristems of A. thaliana according to the model. Floral meristems are initiated on the front of the inflorescence meristem (dashed line). Numbers indicate the developmental stage of floral meristems, as in Figure 5.5(A). The numbers inside the CZ indicate the order of floral primordia, from the youngest (P1) to the oldest (P7)*

Chandler and Werr (2014) used genetic markers to investigate the molecular basis of floral development in *A. thaliana* (Figure 5.6(A)). Expression of the DORNRÖSCHEN-LIKE (DRNL) gene marks all organ founder cells and precedes the auxin response maxima, as measured by the DR5 reporter gene (Chandler et al. 2011).

Effective distances from the center of the meristem (R) are color coded from dotted to black: the shorter the distance, the higher the inhibition intensity.

Modeling the unidirectional initiation of sepals (from the ventral (K1) to the dorsal side (K4)) in *A. thaliana* observed by Chandler and Werr (2014) required initialization by a strong adaxial and a weak abaxial repulsion ($R_{Axis} = 2$; $R_{Bract} = 5$) so that K1 is abaxial. The early stages of *Arabidopsis* development are thus axially symmetrical (Figure 5.6(B)).

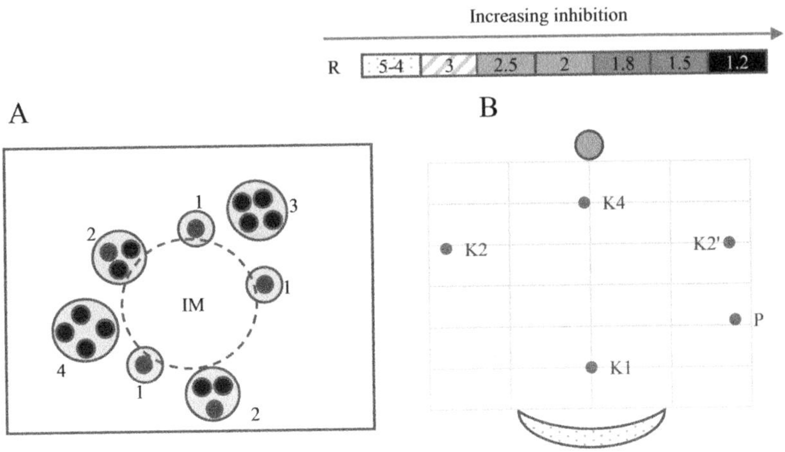

Figure 5.6. *(A) Location of organ primordia on the inflorescence meristem of A. thaliana based on Chandler and Werr's (2014) study. Red primordia correspond to mixed bracts/sepals, and black primordia correspond to sepals. The dashed circle marks the front. Numbers correspond to the developmental stage. (B) Modeling of the A. thaliana floral meristem using Eichler.xls. Primordia are numbered in chronological order (e.g. K1 is the first sepal).* R_{Axis} *= 2 and* R_{Bract} *= 5. Upper circle = inflorescence axis; lower organ = ephemeral bract. K = sepal; P = petal*

Since the 1980s, many studies have focused on the genetic control of *Arabidopsis* flower development, and more particularly on the effects of mutations that disrupt floral morphogenesis. Hepworth et al. (2005) studied the BOP1/BOP2 (BLADE-ON-PETIOLE) genes in *Arabidopsis*. These genes contribute to the control of symmetry, an important aspect of leaf and flower development. Unlike in the wild type, flowers of the *bop1 bop2* double mutant are generally supported by a small bract that appears during the first two stages of floral development. The mature flower is composed of five sepals, four petals, six or seven stamens and two carpels.

To model the bop1 bop2 mutant with dorsal K1, the bract, which does not exist in the wild type, must generate an inhibition stronger than the axis (Figure 5.7(B)).

Curie's (1894) symmetry principle states that "when certain causes produce certain effects, the elements of symmetry of the causes must be found in the effects produced". However, the effects can be more symmetric than the causes (therefore the reciprocal of Curie's principle is not true). Thus, Pierre Curie extended the application of the principle of symmetry used by the crystallographer Auguste Bravais (1850) – who, with his brother Louis Bravais, was also a pioneer in the study of phyllotaxis – to all physical phenomena.

Box 5.2. *Curie's symmetry principle*

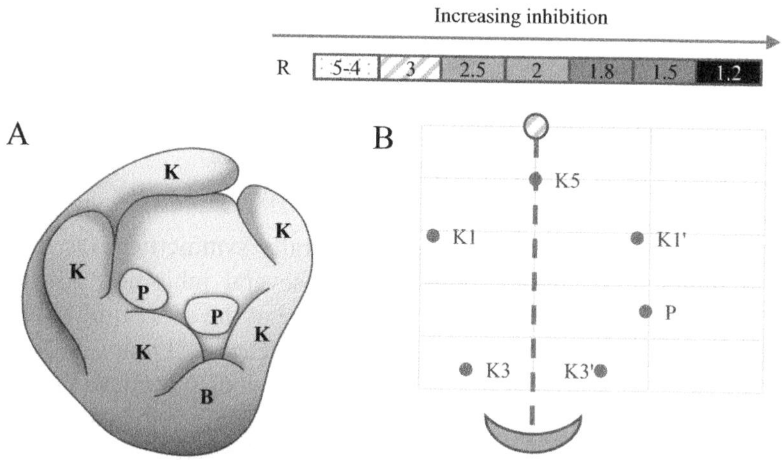

Figure 5.7. *(A) Floral meristem of the bop1 bop2 Arabidopsis double mutant at stage 5: K = sepals, P = petals, B = subtending bract (redrawn from Hepworth et al. (2005)). (B) Modeling of the floral meristem of the bop1 bop2 double mutant using Eichler.xls. Primordia are numbered in chronological order.* R_{Axis} *= 3 and* R_{Bract} *= 2*

The inflorescence axis and the bract (the "causes") are on the vertical line passing through the flower (dashed line in Figure 5.7(B)): consequently, the inhibition potential they produce ("the effect") is bilaterally symmetrical. Two potential minimums (where K1 and K1' are initiated) are therefore bilaterally symmetrical, in accordance with Curie's symmetry principle (Figure 5.8).

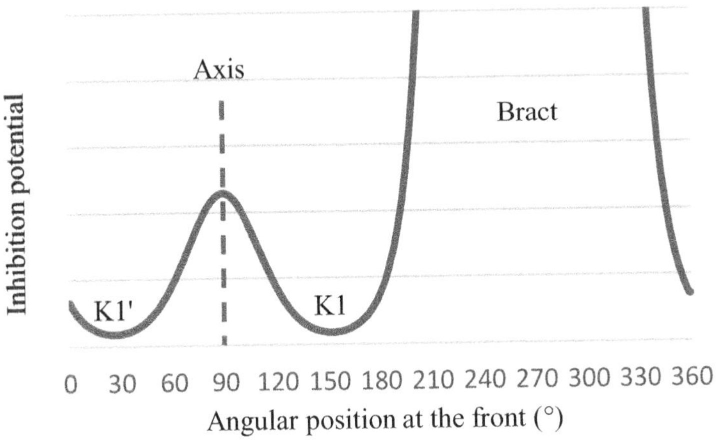

Figure 5.8. *Profile of the inhibition potential on the meristem front at the initiation of the first sepal K1 in the bop1 bop2 double mutant, with R_{Axis} = 3 and R_{Bract} = 2*

When the initial parameters are bilaterally symmetrical (as in the *A. thaliana bop1 bop2* double mutant), several inhibition potential minimums appear with equal potential value at the same time (Figure 5.8), which is not the case for the asymmetrical (Figure 5.4): the sepals (K1 and K1') will be initiated simultaneously because they are indistinguishable. However, our *basic* model being purely recursive, primordia appear one after the other. Therefore, K1 would appear before K1', and thus the inhibition it produces would modify the position of the latter, breaking the initial symmetrical arrangement. We have thus modified the basic model to force primordia (here K1 and K1') to simultaneously occupy the two minimums according to their former angle in case of bilateral symmetry.

When there is bilateral symmetry, use the "Bilateral" tab of Eichler.xls.

Box 5.3. *Free computer software*

In accordance with Curie's symmetry principle, if the sources of inhibition are axial, the primordia must also present bilateral symmetry. Thus, sepals K1 and K1' are dorsal, K3 and K3' ventral and K5 adaxial (Figure 5.7(B)). In the mutant, the fifth organ, which had petal identity in the wild type (P), is located on the adaxial side of the flower and takes on sepal identity. There is therefore a correlation between position and identity: it is difficult to imagine a sepal slipping between K1 and K2' in the wild type (Figure 5.6(B)).

5.3. The history of the contact pressure model of phyllotaxis

In 1878, the Swiss botanist Simon Schwendener proposed a contact pressure model of phyllotaxis, namely that the arrangement of leaves is the consequence of the pressure exerted by each primordium on neighboring primordia.

For Schwendener, the two primordia located on the two parastichies crossing a given primordium (i) produce, by contact, a pressure on this primordium (i). Using geometric methods that were inspired by those used in architecture for gabled roofs formed by sloping rafters of unequal lengths, he believed he had shown that starting from a (1, 2) mode, phyllotaxis converges toward the golden angle. However, his proof was wrong (Adler 1977).

According to Church (1904), this theory only applies to primordia once they are formed. It does not explain the initial arrangement, as evidenced by the existence of meristems presenting phyllotaxic spirals without having primordia in contact with each other (Nymphaea, *Sempervivum*). The rhizome of the fern *Aspidium filix-mas* bears under its scales papillae-like primordia that are not in contact with each other. Their arrangement nevertheless follows Hofmeister's law, and is in a (3, 5) phyllotaxis mode.

But secondary displacements from the site of initiation can be due to the contact pressure and come from three causes:

– Older primordia exert pressure on younger primordia after they are initiated.

– Growing primordia exert reciprocal pressure when there is direct contact.

– Pressure can be the effect of a *rigid boundary* within a growing system.

The influence of rigid boundaries, which is observed in buds of *Cedrus libani*, with a (5, 8) phyllotaxis mode produces the hexagonal shape of the future leaves. So, contact between primordia may cause no overall displacement, but may only affect the edges, which can modify primordium shape.

For Adler (1977), *contact pressure is represented by the hypothesis that the minimum distance between lattice points is maximized*. He showed that if (m, n), with m < n, is the contact phyllotaxis resulting from when contact pressure first becomes effective, then continued contact pressure leads to higher phyllotaxis modes as r decreases, and must follow consecutive terms of the Fibonacci sequence generated by the numbers m and n, namely m, n, m + n, m + 2n, 2m +3n and so on.

Ridley (1982) modeled phyllotaxis as the consequence of a two-step biological process: first, primordia initiation, followed by their interaction during meristem growth. He showed that the divergence angle converged rapidly toward the golden angle, irrespective of the value of the divergence angle between the first primordia. After the initiation of a primordium, the contact pressure exerted by all existing primordia was calculated and produced a rearrangement of these primordia. Primordia movement under the effect of contact pressure was calculated for each primordia pair (provided they were close enough), starting from the youngest to the oldest primordium, as follows: primordium (i) moves away from primordium (j) by a vector whose direction is that of the axis (i → j) and by a length that is less than the distance d_{ij}. Several iterations of this algorithm were performed before allowing the initiation of a new primordium.

In 2006, Hellwig et al. considered that models based on chemical inhibitors were "purely hypothetical" insofar as they only yielded satisfactory results from initial conditions that cannot be verified in real plants. By contrast, models of the effect of contact pressure, provided primordia are initiated between 127° and 149°, move primordia to a position where they are as far apart as possible from each other and create Fibonacci spirals of high modes. Even better results were obtained when new primordia were initiated according to Hofmeister's rule.

Researchers from Osaka University (Kitazawa and Fujimoto 2015) also assumed two stages: (1) an initiation stage where each primordium arises successively at the least crowded position, depending on the inhibition potential function U_{ini}, and (2) a growth stage where the arrangement is determined by repulsion among primordia. They assumed that, during this stage, all primordia repel each other.

The repulsion exerted on primordium k is quantified by an exponentially decaying potential. When there are n primordia ($1 \leq k \leq n$)

$$Ug_k(R,\theta) = \sum_{j=1, j\neq k}^{n} \exp(-\frac{d_{kj}}{\delta}) \qquad [5.1]$$

The primordia descend along the gradient of the inhibition potential Ug to find a location with weaker repulsion. This formulation, which corrects the divergence angle after initiation, is similar to the contact pressure model.

5.4. Modeling the floral meristem of *Illicium*

Austrobaileyales is an order of flowering plants comprising approximately 100 species of woody plants growing as trees, shrubs and lianas. The best-known species is *Illicium verum* (Schisandraceae), commonly known as star anise. The order belongs to the ANA grade of basal angiosperms (Amborellales, Nymphaeales and Austrobaileyales), which diverged from the lineage early on, leading to all other extant flowering plants. This order comprises three families: Austrobaileyaceae, consisting of a single species, *Austrobaileya scandens* (a woody liana), Schisandraceae, a family of trees, shrubs and lianas which is the basis of essential oils, and Trimeniaceae, trees and lianas which are also the basis of essential oils.

At early stages of floral development, all organs of *Illicium anisatum* are spirally arranged, the divergence angles being near 138° (Endress 2001; Figure 5.9(A)). Phyllotaxis is in a (3, 5) mode. Eight carpels are located on the meristem front not already formed, as shown in Figure 5.9(A).

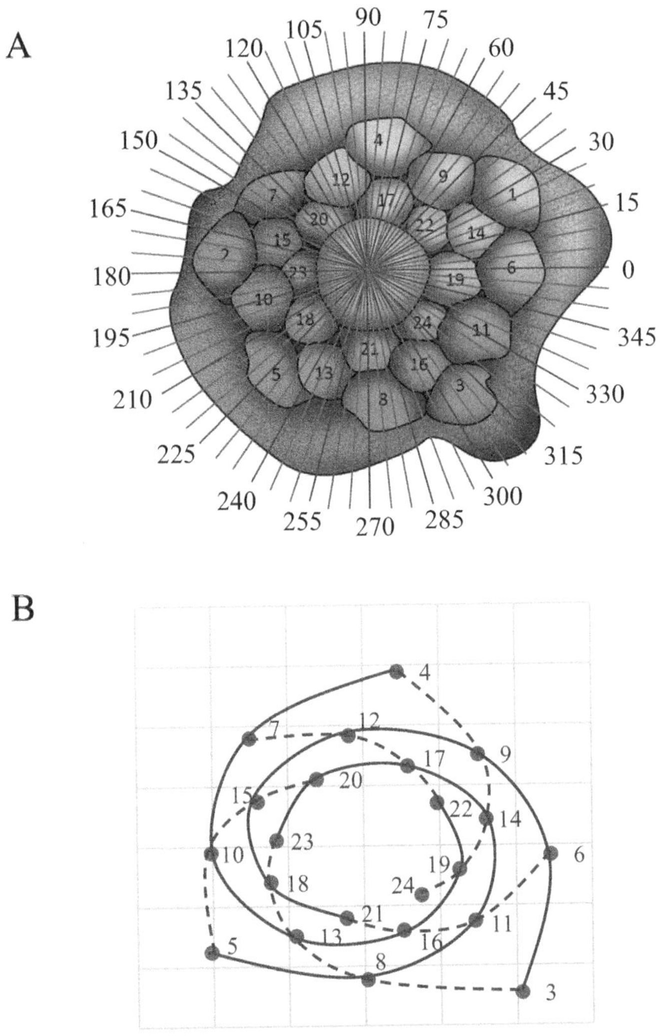

Figure 5.9. *(A) Floral organ initiation in Illicium anisatum (redrawn from a SEM micrograph from Endress (2001)), with a protractor centered on the apex of the FM. (B) Model of the floral meristem of Illicium anisatum*

We used the dynamical model (equations [4.1] and [4.2]) to simulate *Illicium* phyllotaxis and Table 1.1 data for parametrization (Walch and Blaise 2022a).

To compare the results of the model with observational data (Figure 5.9(A)), we calculated the angular position of the center of each primordium, defined as the half sum of the angular positions of the two edges of the primordium measured to the nearest degree with a protractor on the SEM micrographs from Endress (2001). We then calculated the root-mean-square error (RMSE) between the observed *angular positions* and the angular positions derived from the model, and compared it to the mean angular size of the primordia (Table 5.1). The radial position of the primordia is a parameter of the model.

Primordium number	RMSE (°)	Mean size (°)
1-21	8	37

Table 5.1. *Root-mean-square error between the observed and modeled angular positions of primordia in Illicium anisatum compared to the mean angular size*

The RMSE between the observed and modeled angular data is significantly smaller than the mean primordium size, showing an excellent fit of our model to the observations of the floral meristem of *I. anisatum*.

The root-mean-square deviation (RMSD), or root-mean-square error (RMSE), is a standard measure of the difference between observed and predicted values.

$$RMSE = \sqrt{\frac{\sum_{i=1}^{i=n}(\theta th_i - \theta obs_i)^2}{n}},$$

where θth is the theoretical value calculated by the model and θobs is the observed value for n values

Box 5.4. *Root-mean-square error (RMSE)*

Weisse (1894) had already observed that some species with whorled phyllotaxis at anthesis display a spiral arrangement during development. We modeled this transformation of the gynoecium in *I. anisatum* by simulating a pressure exerted on organ *i* by (n – 1) organs, which, after being initiated in a spiral, are approximately located on the same circle (*the meristem front that will be the future whorl*), the center of which is the meristem apex, using the repulsion potential

$$V_i(R_i, \theta_i) = \sum_{j=1, j\neq i}^{n} \exp\left(-\frac{d_{ij}}{\delta}\right) \quad [5.2]$$

where (R_i, θ_i) are the polar coordinates of organ i, and δ is an exponential decay length different from γ, that is, the exponential decay for primordium initiation in equation [4.2].

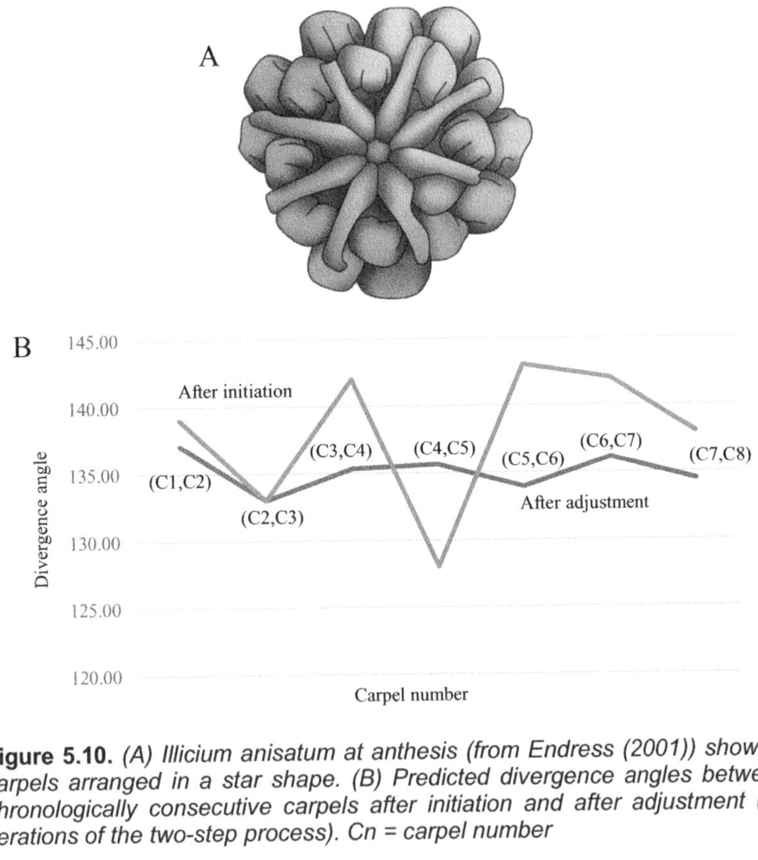

Figure 5.10. *(A) Illicium anisatum at anthesis (from Endress (2001)) showing carpels arranged in a star shape. (B) Predicted divergence angles between chronologically consecutive carpels after initiation and after adjustment (20 iterations of the two-step process). Cn = carpel number*

Primordia descend along the gradient of this potential, that is to say they move from a position where the potential is high to a point where the potential is lower. To calculate the angular potential gradient, we compared the potential at a virtual primordium i' (R, $\theta+d\theta$) and at i (R, θ). The difference in the value of the potential between the real organ and this virtual organ, whose angular position is one degree more, was calculated and used to reposition the n organs by making them go down the potential. From their new positions, we recalculated the gradients of these potentials and obtained new positions, following a *recursive* process.

If a primordium is closer to one of its two neighbors on the circle, then the "repulsion" potential will be higher on that side than on the other, which brings this primordium closer to its more distant neighbor. Ultimately, in a perfect whorled configuration, the potential gradient is zero on both sides of the primordium, the distance from the two neighbors being equal, which ensures the stability of this arrangement.

In our simulation of floral organ initiation in *Illicium*, all primordia were moved at each iteration, but only by a small angular distance that depended on a sufficiently small parameter to ensure that the process converged.

Sokoloff et al. (2018) have insisted that the gynoecium of *Illicum* is spiral in the early developmental stages and whorled at anthesis. Sauquet et al. (2018) contested this interpretation (which is based on the top view of the mature flower; see Figure 5.10(A)) and suggested that the carpels were not perfectly symmetrical. However, phyllotaxis does change after all organs have been initiatied, that is, when all floral organs are present and the position of organs is "adjusted" later in development (Sokoloff et al. 2018).

The contact pressure model explains the transition from a spiral to a whorled arrangement of the carpels of *I. anisatum*. At the end of the recursive process, the angular distances between the primordia on the front are more or less equal (Figure 5.10(B)). The divergence angles between chronologically consecutive carpels, which are unequal after spiral initiation, all approached 135°, which means that two geometrically consecutive carpels in this whorl are separated by 45°, corresponding to a perfectly whorled arrangement. The small residual irregularities justify the reservations expressed by Sauquet et al. (2018) regarding the symmetry of the gynoecium, which is not perfectly radial. Furthermore, traces of spiral initiation are still visible in the aestivation of the carpel flanks.

5.5. Mechanical forces in floral development

Bull-Hereñu et al. (2022) listed many cases where mechanical forces can shape the developing meristem to such an extent that it can be considered part of the ontogenetic process of the plant. For example, if the bracts within an inflorescence are rigid, they can reduce the space available for the axillary meristem and compress the developing flower. In many *monosymmetric (bilaterally symmetrical)* flowers, the sequence of organ initiation tends to be unidirectional, often from the abaxial to the adaxial side (e.g. many Fabaceae, Lamiales), and this is probably related to the fact that internal pressures are stronger from the inflorescence axis (the adaxial side).

In *Baptisia australis* (Papilionoideae) whose inflorescence is a raceme, two bracteoles appear on either side of the axis of the meristem before the initiation of the first sepal, then completely disappear during floral development. They are therefore no longer visible at anthesis (Prenner 2004). The initiation of the sepals is "unidirectional": K1 (ventral) -> K2, K2' (lateral) -> K4, K4' (dorsal) (Figures 5.11(A) and (B)).

According to our model, the two upper bracts in the inflorescence (Figure 5.11(A)) are necessary to reinforce the inhibition produced by the two very small bracteoles and produce a unidirectional floral organ initiation pattern (absolute angles: $\theta_{B\text{-}3} = 80°$; $\theta_{B\text{-}5} = 100°$; $\theta_{b1} = 150°$; $\theta_{b2} = 30°$; Figure 5.11(C)). Indeed, they push the sepal K1 abaxially and K2 away from the axis. These bracts therefore influence the arrangement of the flower, according to Bull-Hereñu et al.'s (2022) interpretation of unidirectional initiation. Moreover, they slow down the axial development of the floral meristem, which gives it its elliptical shape.

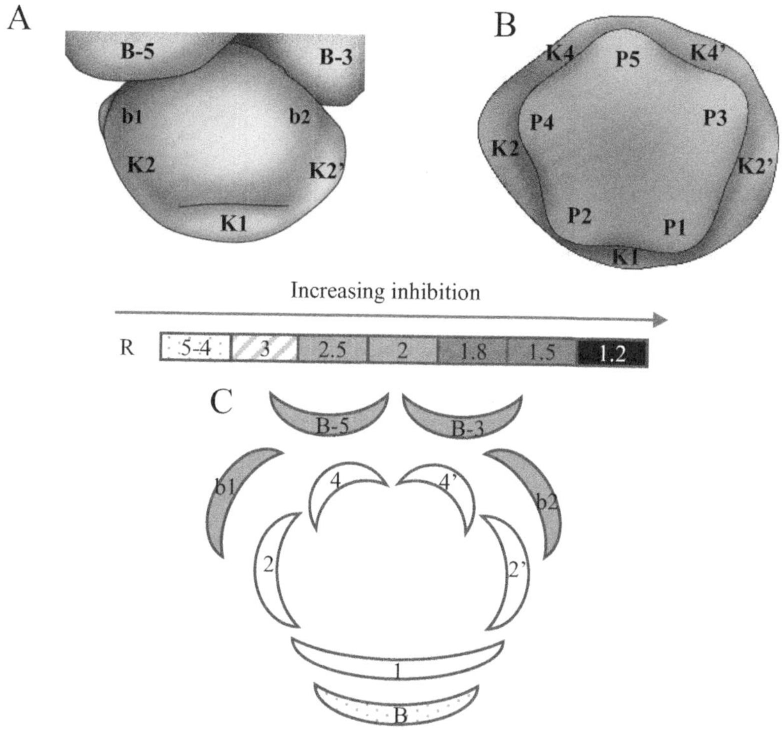

Figure 5.11. *(A) Initiation of the bracteoles and the perianth in Baptisia australis (Papilionoideae). The first sepal is initiated in the abaxial position (redrawn from Prenner (2004)). (B) The bracteoles have disappeared (Prenner 2004). (C) Model of the unidirectional initiation of floral organs in Baptisia australis. b = bracteoles; K = sepal; P = petal, B = bract. The sickles B_{-3} and B_{-5} have been reversed because they are not bracteoles but bracts of the upper flowers*

5.6. Physical models

Rothen and Koch (1989), from the Institute of Experimental Physics at the University of Lausanne, "idealized" phyllotaxis as a lattice in which the genetic spiral is logarithmic, that is, the plastochron ratio is constant and elements are circles that are tangent to each of their neighbors (Figure 5.12).

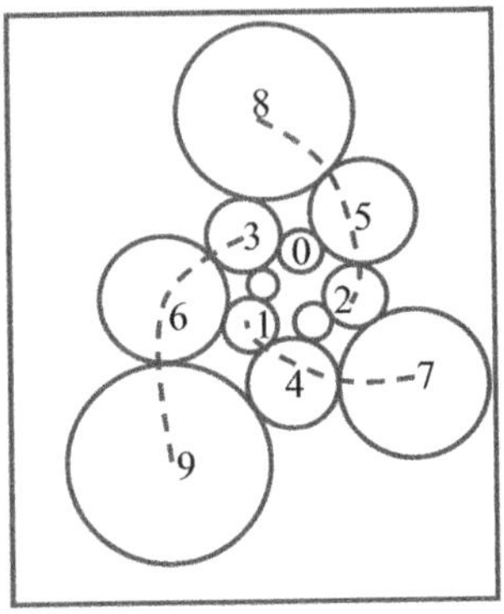

Figure 5.12. *Lattice of circles that are tangent to each other and whose centers lie on a logarithmic spiral according to Rothen and Koch (1989). The lines joining the contact points are the parastichies*

They demonstrated that the number of parastichies belongs to the Fibonacci sequence and that when the plastochron ratio is decreased, the system rises in the Fibonacci sequence by a succession of "quasi-bifurcations". Indeed, at each bifurcation, such as from (5, 8) to (8, 13), the passage to (5, 13) is impossible, because these two types of parastichies have the same chirality (direction of rotation). More generally, it is congruent with van Iterson's diagram, where each branch tends toward a noble angle.

Levitov (1991a, 1991b), from the Landau Institute for Theoretical Physics in Moscow, discovered that structures that are similar to plant phyllotaxis appear in layered superconducting systems, that is, systems that offer no resistance to the passage of electricity when subjected to temperatures close to absolute zero and to a magnetic field. He concluded that phyllotaxis is a general phenomenon that can occur in all "soft lattice" subjected to strong deformations.

Levitov showed that in layered superconducting systems, the minimums of the repulsion energy follow quasi-bifurcations that correspond to van Iterson's diagram, where the magnetic field to which the superconductor is subjected increases.

By applying the mathematical theory of modular groups to lattices of mutually repulsed soft objects subjected to strong deformation, Levitov found patterns that corresponded to van Iterson's diagram. These trajectories apply to botanical phyllotaxis, but also to cells of natural convection (Rayleigh-Bénard) and flux lattices in superconductors. Trajectories represent the position of energy minimums under continuous pressure variation. At each bifurcation, only one branch can be followed by the system, which makes it a "quasi-bifurcation".

Douady's (1996a) experimental apparatus consisted of a dish filled with silicone oil, placed in a magnetic field where the intensity is minimum at the center and maximal at the periphery. Drops of magnetic ferrofluid particles simulate biological primordia. A drop that falls in the center of the dish is attracted toward the region of high field. Drops repel each other, therefore existing drops force a new one to move to the largest available space at the periphery of the central region.

Douady and Couder (1996b) set an inhibition potential threshold U_0 for the appearance of a new primordium (it can only appear if the potential is less than U_0). They used several types of energy laws:

$$U(D) = \frac{1}{\left(\frac{D}{D_0}\right)^p} \qquad [5.3]$$

where D is the distance to the generating primordium of size D_0.

According to Green (1999), if a ring is confined within a circular region of fixed radius and is "forced to grow", the excess length produces an out-of-plane undulation of five wavelengths. The shape of this undulation depends on the ratio of the elasticity of the inner cells (the corpus) to the flexural rigidity of the epithelium (the tunica).

Newell et al. (2008) proposed a continuum description of the biochemical reaction-transport model. They showed that the partial differential equation governing this biochemical process is almost identical in form to that from the buckling model, and indeed to that governing the behavior of a universal class of pattern-forming systems. They suggest that both biochemical processes (auxin localization) and mechanical forces (non-uniform stress distribution) may *together* play key roles in phyllotaxis. Their model only applies to the initiation of primordia in the generative region.

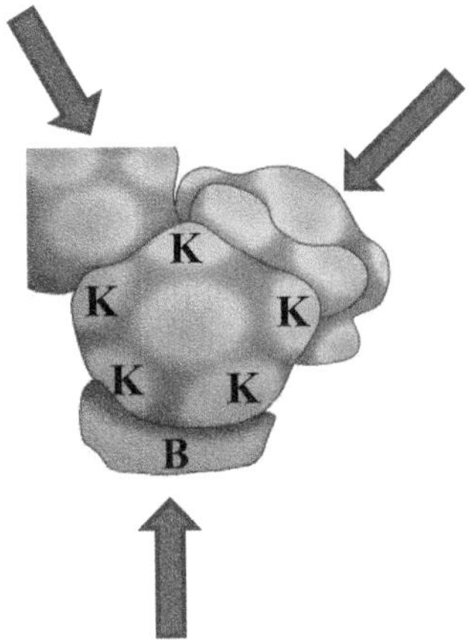

Figure 5.13. *The formation of the pentamerous perianth of Neptunia pubescens (Mimosoid clade) could be the consequence of buckling under the effect of constraints (arrows) on the growth of the meristem (redrawn from a SEM micrograph from Tucker (2003))*

This generative region, where primordia are formed, is an annular region near the meristem apex. Basically, the tunica in the generative region cannot expand in the circumferential direction. A compressive stress may cause the plant surface to buckle. In addition, variation in growth as a result of unequal auxin distribution also leads to local stress changes. These two processes can be strengthened by nonlinear resonance, which allows the formation of polygons of various shapes.

6

Magnoliales and Laurales

6.1. Stability of the Fibonacci spiral

Using the disk stacking model, Golé et al. (2016) showed that a rapid decrease in disk diameter led to the breakdown of Fibonacci spirals in favor of a "quasi-symmetric" arrangement, with a nearly equal number of clockwise and anti-clockwise spirals. In a follow-up paper, Douady and Golé (2016) presented real life examples of botanical structures displaying this pattern.

In our model, the plastochron ratios measured in the vegetative meristem of *L. usitatissimum* (Williams 1975) (Table 1.1) that produce a Fibonacci pattern can be approximated with a decreasing exponential function of organ number (n):

$$r(n) = r_0 e^{-\rho n} + r_{asympt} \quad [6.1]$$

with $r_0 = 0.35$, $\rho = 0.18$ (the decrease rate) and $r_{asympt} = 1.05$ (the level of the asymptote) (Figure 6.1(A)). Starting with two opposite primordia, we obtained with these parameters a rapid convergence of the divergence angle toward the golden angle (Figure 6.1(B)) as well as a (5, 8) phyllotaxis. We call these parameters "the canonical parameters": Fibonacci phyllotaxis requires a "slow and controlled" decrease in plastochron ratios.

For a color version of all the figures in this book, see www.iste.co.uk/walch/phyllotaxis.zip.

This pattern corresponds to the floral arrangement of *Magnolia soulangeana* (Figure 6.9).

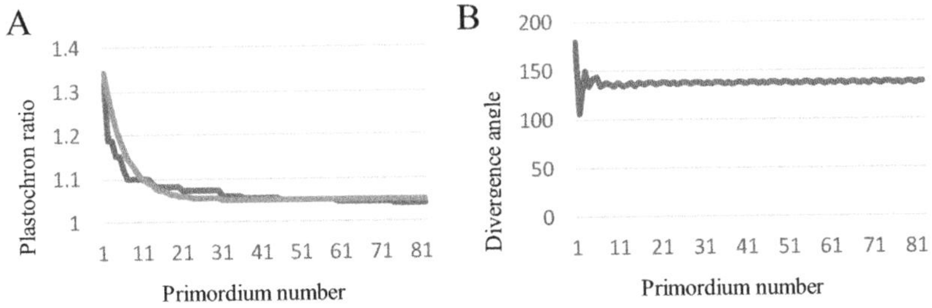

Figure 6.1. *(A) Approximation of the plastochron ratios of leaf primordia in L. usitatissimum by a decreasing exponential function (in orange) with r_0 = 0.35, ρ = 0.18 and r_{asympt} = 1.05 superimposed on the measurements made by Williams (1975) (in blue). (B) Convergence of the divergence angle in the canonical model*

There are two ways the plastochron ratio can be slowed down: by lowering the level of the asymptote r_{asympt} or by increasing the rate of exponential decrease ρ. If we lower the asymptotic value, phyllotaxis will climb the Fibonacci sequence and consequently present a higher number of spirals; for example, according to the phyllotaxis frieze (Figure 4.4), phyllotaxis goes up to (13, 8) when r = 1.02. In flowers, the asymptotic value of the plastochron ratio is generally equal to the plastochron ratio of the last stamens. For r = 1.02, they are very small.

This pattern is that of androecium of *Liriodendron tulipifera* (Figure 6.13).

In addition to lowering the asymptotic value, if we increase ρ (the decrease rate), we find that after the initiation of a certain number of organs, *organ permutations* occur: the divergence angles oscillate periodically between 137°, 223° (= 360° – 137°) and 274° (= 2 × 137°). A primordium is placed at 2 × 137° = 274° instead of at 137° (the tip of the "M" on Figure 6.2(A)), the next primordium is at 223° (the hollow of the "M" on the Figure), which is the reverse direction of 137° and corresponds to where the first organ should have been, the next primordium is at 2 × 137° (the second tip of the "M" on the figure) to skip the place occupied by the first organ (Figure 6.2(C)). Nevertheless, we still observe Fibonacci parastichies in both directions on the cylindrical representation of organ arrangement (Figure 6.2(B)).

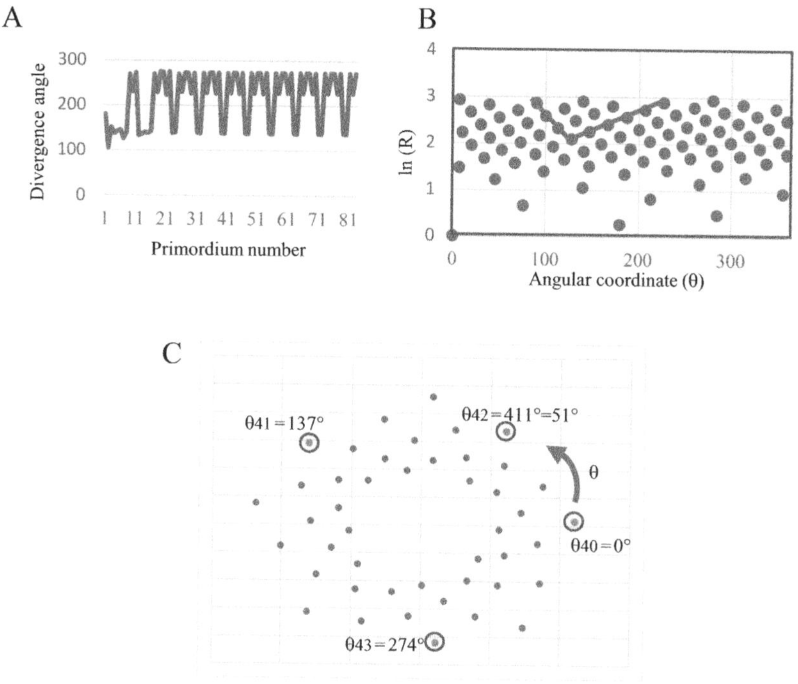

Figure 6.2. *Pattern obtained with r_0 = 0.35, ρ = 0.22 and r_{asympt} = 1.02: (A) Divergence angle showing M-shape. (B) Cylindrical representation of organ arrangement: the angular coordinates (θ) of the organs are on the x-axis and the natural logarithm of the radial coordinates on the y-axis. Parastichies are represented by inclined line segments. (C) Centric representation of organ arrangement*

In this pattern, like in flowers with Fibonacci phyllotaxis, the number of left and right parastichies differs widely (the ratio of these numbers is close to the golden ratio = 8/5 = 1.6), which results in very different inclinations of these two types of parastichies on a cylindrical representation (Figure 6.2(B)).

Divergence angles should be ~137°, for instance, between primordia 40 and 41, as shown in Figure 6.2(C) (ontogenetic spiral is anti-clockwise). However, because there is permutation between primordia 43 and 42, the location of primordium 42 is 137 + 274 = 411°, and of 43 is 411 – 137 = 274°.

These periodic permutations giving divergence angles that follow an *M-shape* have also been observed by Godin et al. (2020) with the disk stacking model. When disks become smaller, they induce small irregularities.

This pattern is that of androecium and gynoecium of *Doryphora aromatica* (Figure 6.21).

When ρ is very high (ρ = 0.7), the divergence angles become chaotic (Figure 6.3(A)).

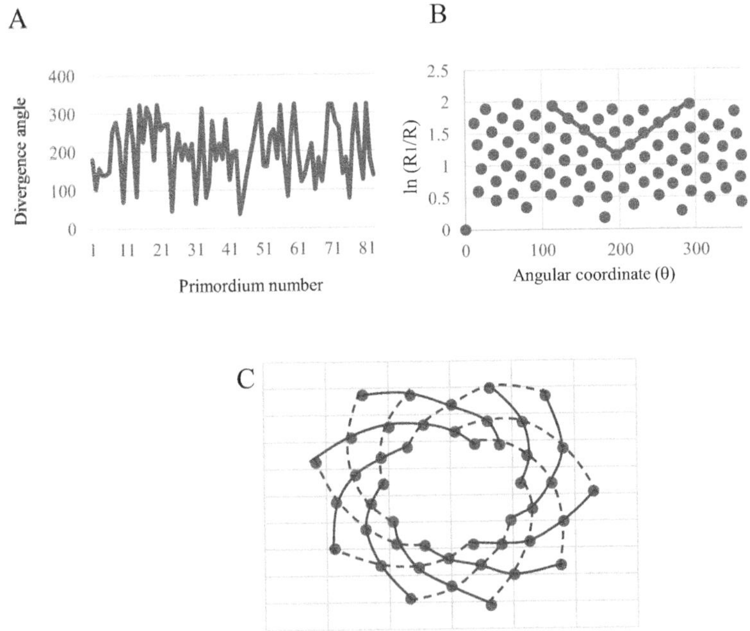

Figure 6.3. *Pattern obtained with r_0 = 0.35, ρ = 0.7 and r_{asympt} = 1.02: (A) Divergence angles. (B) Cylindrical and (C) centric representation of organ arrangement*

For $r_{asympt} = 1.02$ and $\rho = 0.7$, we find nine pseudo-parastichies in both directions ((9, 9) mode) (Figure 6.3(C)). The inclination of the parastichies in both directions is more or less equal (Figure 6.3(B)). Thus, phyllotaxis is *quasi-symmetric*. This type of arrangement is observed in the stamens of Nymphaeaceae and has been interpreted as "complex-whorled" (i.e. non-spiral) by Endress and Doyle (2015) (see also El et al. 2020; Walch and Blaise 2022a). This is the pattern of *Magnolia denudata* and *Annona hexapetala* (Figure 6.17).

6.2. Transient regime

Douady and Couder (1996c) studied the transient regime from a distichous to a spiral mode. Thus, in a seedling of *Arabidopsis*, starting with two opposite cotyledons, the next primordia form at a right angle to the cotyledons, as in a decussate mode, but *do not form simultaneously*. This breaks the initial symmetry, and therefore the next primordium is not at a right angle to the previous two primordia but is closer to the older one. The Douady and Couder model fits this pattern.

To model the transition from decussate to spiral phyllotaxis, we have added an additional property to the original Douady and Couder model, namely that the potential generated by each primordium can decrease exponentially with the age of the primordium generating the field, allowing the decay rate α_i to be different between primordia (as in Kitazawa and Fujimoto 2015):

$$U_i(\theta) = \sum_{j=1}^{i-1} \exp\big(-\alpha_i(i-j-1)\big)\exp\left(-\frac{d_{ij}}{\gamma}\right) \qquad [6.2]$$

To produce a Fibonacci spiral, if we start with two cotyledons at 180° from each other, we must apply to these two primordia a rate of decrease in their potential with age of $\alpha_1 = \alpha_2 = 1$, all other α_i being equal to 0. Thus, when the third primordium is initiated, the inhibition from the older primordium is weaker than that of the younger. The initial symmetry is broken, and thus phyllotaxis definitely changes from distichious to spiral.

If the plastochron ratio is constant, the distance of a primordium to the center of the meristem at time t is:

$$R = R_0 \exp(\frac{V_0 t}{R_0}) \quad [6.3]$$

where V_0 is the speed of radial displacement.

The plastochron ratio is equal to:

$$r = \frac{R_{n-1}}{R_n} = \exp(\frac{V_0 T}{R_0}) \quad [6.4]$$

where T is the plastochron (the delay between the initiation of two successive primordia).

The growth index G is (see Chapter 2):

$$G = ln(r) = \frac{V_0 T}{R_0} \quad [6.5]$$

Box 6.1. *Speed of the radial displacement of primordia (Douady and Couder 1996a)*

6.3. Continuity of the Fibonacci spiral

Meicenheimer (1979) described the development of the shoot meristem from the vegetative stage up to the initiation of all floral organs in *Ranunculus repens* showing the quasi-continuity of the generative spiral in this species.

In *R. repens*, the phyllotaxis mode rises from (2, 3) during the vegetative stage to (3, 5) during the transitional stage (the formation of bracts and bracteoles) to (5, 8) during sepal initiation, all these patterns corresponding to a Fibonacci spiral. Petals, however, are initiated in a pentamerous whorl. Spiral phyllotaxis resumes with the initiation of stamens and carpels in a (8, 13) mode. The corolla is the only discontinuity within this spiral (Table 6.1).

Meicenheimer measured the plastochron ratio at each developmental stage: plastochron ratios decrease during the developmental sequence, resulting in the rise in phyllotaxis modes. These data were confirmed by

Rutishauser (1998). These values allowed us to parameterize our model, which faithfully reproduced the floral diagram of *R. repens* (Figure 6.4). The divergence angle between the two bracteoles was set at 144°, in accordance with Meicenheimer's data, which here makes the bracteoles the transmission belt between vegetative and reproductive phyllotaxis, and corresponds to their role in determining the arrangement of the sepals (see Chapter 7).

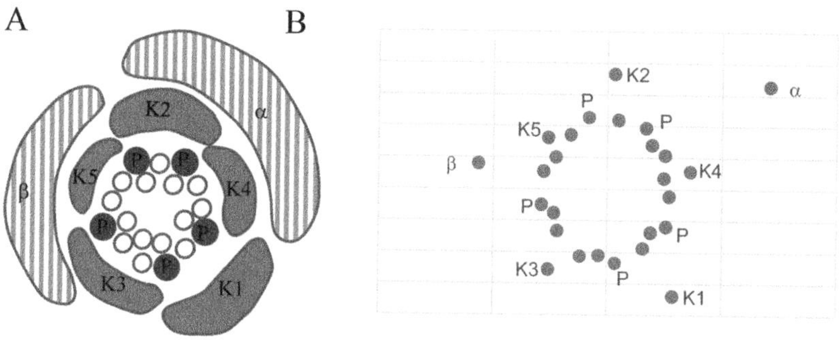

Figure 6.4. *(A) Floral diagram of Ranunculus repens, redrawn from Meicenheimer (1979). (B) Model of the floral meristem of R. repens using the plastochron ratios measured by Meicenheimer as parameter values. α, β = bracteoles; K = sepals; P = petals. Unlabeled circles = stamens*

Stage	Floral primordium number	Mean plastochron ratio	Fibonacci mode
Vegetative		1.42	(2, 3)
Transitional (bracteole)		1.20	(3, 5)
Sepal	1–5	1.116	(5, 8)
Petal	6–10	1.000	
Stamen	11–23	1.012	(8, 13)
Carpel		1.017	(8, 13)

Table 6.1. *Mean plastochron ratios of consecutive floral organ primordia in Ranunculus repens measured by Meicenheimer (1979)*

The average divergence angle between sepals is 140°, with values ranging between 131° and 153°.

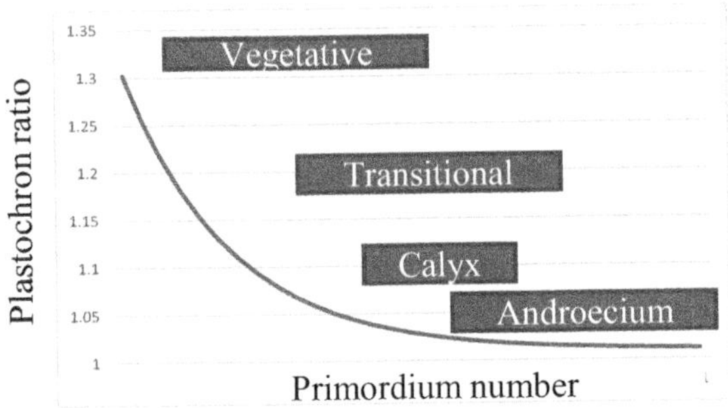

Figure 6.5. *Continuity of the Fibonacci spiral in the shoot meristem of Ranunculus repens according to Meicenheimer (1979) and our model (blue curve) with plastochron ratio according to canonical parametrization*

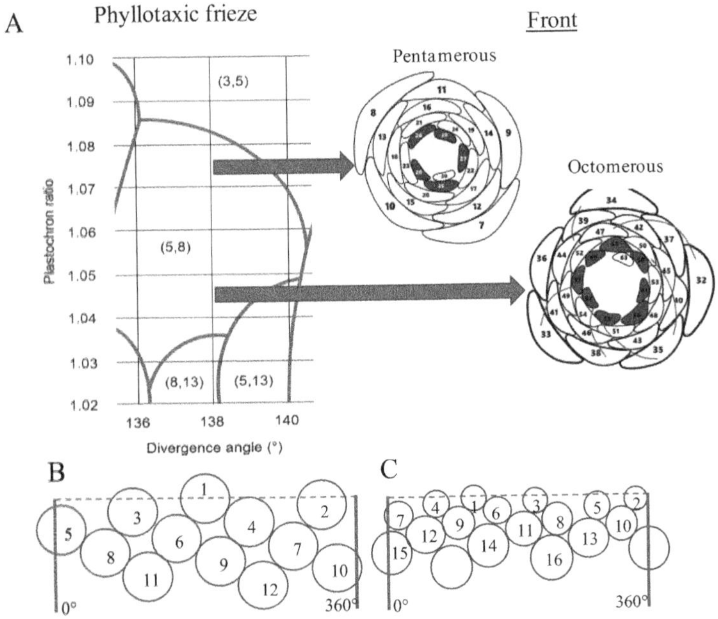

Figure 6.6. *(A) Front for different plastochron ratios in centric representation according to Linum usitatissimum data. (B and C) Cylindrical representation of trimerous and pentamerous fronts (dashed lines) from Godin et al. (2020)*

All vegetative and floral organs are initiated on the same Fibonacci spiral with the exception of the corolla. According to our model with the canonical parametrization (Figure 6.1(A)), this results in a regular decrease in the plastochron ratio (Figure 6.5), and thus, as seen from the phyllotaxic frieze (see Chapter 4), in a rise in Fibonacci modes (Figure 6.6(A)). In the (5, 8) mode, corresponding to the formation of the calyx, the front where the sepals are initiated is made up of five primordia (Figures 6.6(A) and (C)).

Schimper and Braun (1835) and Bravais and Bravais (1837a) had already noticed that the generative spiral extended from the leaves to the floral organs in many Ranunculaceae (*Ranunculus*, *Adonis*, *Helleborus*, *Nigella*, *Caltha*, *Trollius*), and also in *Magnolia* and *Liriodendron*.

6.4. Magnoliales

The order Magnoliales contains six highly diverse families. Flowers are usually composed of a weakly differentiated perianth arranged in three trimerous whorls and a spirally arranged androecium. For a long time, *Magnoliaceae* were considered ancestral within angiosperms. But in the 1990s, the phylogeny of flowering plants was reviewed in the light of genetic analyses (Chase et al. 1993). The orders Amborellales, Nymphaeales and Austrobaileyales were found to form a grade (the ANA grade) at the base of the angiosperm tree (APG IV 2016; Ronse de Craene 2022; Walch and Blaise 2022a).

The flowers of *Annonaceae* are usually trimerous with three perianth whorls. The androecium can consist of multiples of three stamens. The six outer stamens (or staminodes, that is, stamen-like sterile organs) tend to be inserted as three pairs in alternation with the tepals. The androecium can also be highly polymerous (up to 2,000 stamens in *Annona montana*) with a more or less whorled initiation, becoming spiral higher up on the receptacle (Ronse de Craene 2022).

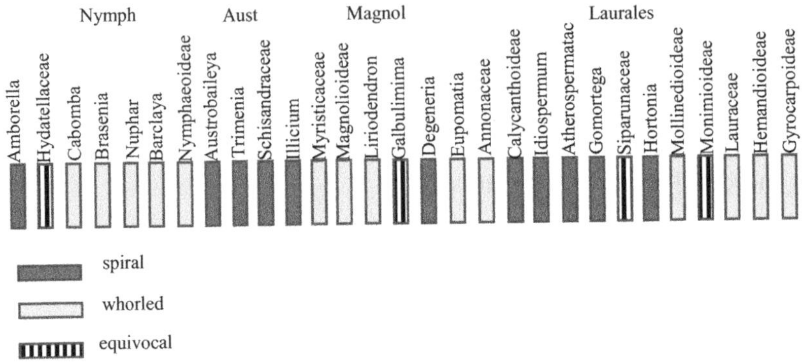

Figure 6.7. *The perianth phyllotaxis of basal angiosperms, Magnoliales and Laurales according to Endress and Doyle (2015). Nymph = Nymphaeales; Aust = Austrobaileyales; Magnol = Magnoliales*

Endress and Doyle (2015) mapped the phyllotaxis of the perianth of basal angiosperms (ANA grade), Magnoliales and Laurales to the phylogenetic tree (Figure 6.7).

6.4.1. *Magnoliaceae*

Magnoliaceae are a family of more than 200 species of trees and shrubs with large flowers, most of which are found in the subtropical and tropical regions of East Asia. There are two recognized genera in this family: *Magnolia* (magnolia) and *Liriodendron* (tulip tree). In Magnoliaceae, the transition between the large perianth parts and the androecium is abrupt and leads to a disruption of spiral phyllotaxis. The initiation of the androecium does not follow a regular spiral: the outer stamens tend to arise in flushes opposite the inner tepal whorl (Ronse de Craene 2022).

Flowers of Magnoliaceae are often enclosed in a bag-shaped spathe which unfolds at anthesis, and which can be considered a bract or bracteole: it is located just above the last leaves, which exhibit Fibonacci phyllotaxis (Eichler 1878) (Figure 6.8). Here, we consider that the spathe acts as a transmission belt between vegetative and reproductive phyllotaxis.

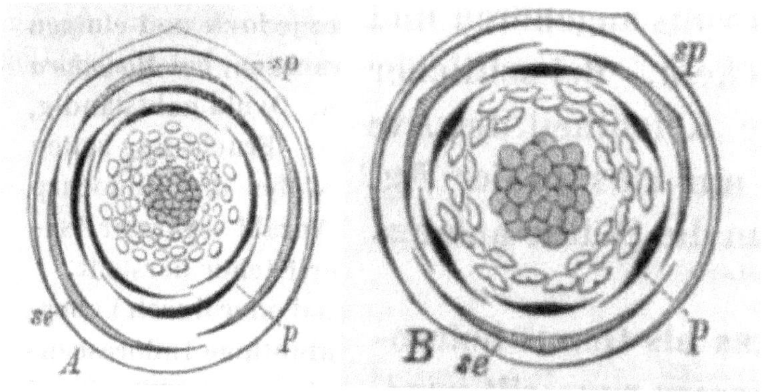

Figure 6.8. *Floral diagrams of (A) Magnolia grandiflora and (B) Liriodendron tulipiflora (from Eichler 1878). sp = spathe; se = sepal; P = petal*

6.4.1.2. *Magnolia*

Wiss and Zagórska-Marek (2012) have synthesized decades of research on phyllotaxic patterns in *Magnolia salicifolia, M. stellata, M. denudata* and *M. acuminata*. The perianth parts are generally arranged in three trimerous whorls, initiated as large primordia within relatively small meristems. The transition from tepals to stamens is accompanied by an increase in the size of the meristem and a decrease in primordia size. Small primordia sizes as well as the many local changes in phyllotaxis explain the great diversity of spirals observed in the androecium.

Several dozen different phyllotaxis patterns in the androecium have been observed in different species, some of which are common, for example, Fibonacci (8, 13), bijugate (6, 10) or Lucas (7, 11) phyllotaxis, whereas others are rare. A change in phyllotaxis pattern can occur with the disappearance or appearance of a spiral, such as (8, 11) → (7, 11) or (8, 12) → (8, 13). They can be the consequence of local variations in the size of the meristem or the primordia (Wiss and Zagórska-Marek 2012). The phyllotaxis modes observed in the gynoecium are lower than those of the androecium, in agreement with our phyllotaxic frieze, given the increased ratio of primordia to meristem size in the gynoecium.

The floral apex of *Magnolia soulangeana* shows a quite regular Fibonacci pattern with a (5, 8) mode (Zagórska-Marek 1994) (Figure 6.9(A)).

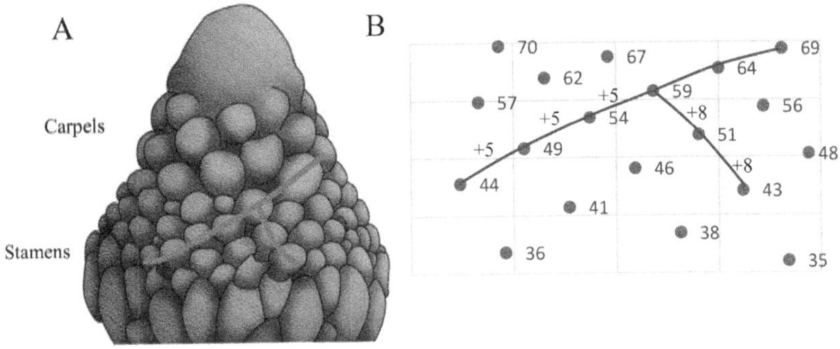

Figure 6.9. *(A) The floral apex of Magnolia soulangeana during stamen and carpel initiation, redrawn from Zagórska-Marek (1994). Two partial parastichies are shown. (B) A cylindrical representation of floral organ initiation in M. soulangeana modeled using the canonical parameters:* $r_0 = 0.35$*,* $\rho = 0.18$ *and* $r_{asympt} = 1.03$*. The same parastichies as in A are shown*

To model the initiation of floral organs in *M. soulangeana*, we used the canonical plastochron ratios. To initialize the model, the first sepal was considered to be the fifth organ on the same ontogenetic spiral as the vegetative spiral (Table 6.2). This transition from the vegetative stage to the reproductive stage gives the perianth a lower plastochron ratio.

In flowers with Fibonacci phyllotaxis, the number of left and right parastichies differs widely (the ratio of these numbers is close to the golden ratio ~8/5 = 1.6), which results in very different inclinations of these two types of parastichies on a cylindrical representation (Figure 6.9(B)). The difference in primordium number along the parastichies follows the (5, 8) mode: for instance, from primordium 44, the next primordium on one of the five spirals is primordium 49 (44 + 5) and from primordium 43, the next primordium on one of the eight other spirals is primordium 51 (43 + 8) (Figure 6.9(B)).

Primordium number	Organ	Plastochron ratio
1	Vegetative	1.32
2	Vegetative	1.27
3	Vegetative	1.23
4	Vegetative	1.20
5	K1	1.17
6	K2	1.15
7	K3	1.13
8	K4	1.11
9	K5	1.10
10	K6	1.09
11	K7	1.08
12	K8	1.07
13	K9	1.06
14	Stamen	1.06
15	Stamen	1.05
16	Stamen	1.05
17	Stamen	1.05
18	Stamen	1.04
–	–	–

Table 6.2. *Plastochron ratios used to model floral organ initiation in Magnolia soulangeana*

Magnolia denudata is a shrub native to central and eastern China. It has been cultivated since the Tang Dynasty for the beauty of its large, white, fragrant flowers.

According to Erbar and Leins (1994), its trimerous perianth represents a transition between spiral and whorled phyllotaxis. The first three tepals are separated by short plastochrons, followed by a long plastochron marking the transition to the second "whorl". The next two plastochrons are short, followed by a longer plastochron separating the second "whorl" from the third "whorl". Erbar and Leins called this pattern "rhythmisation" of spiral to whorled phyllotaxis.

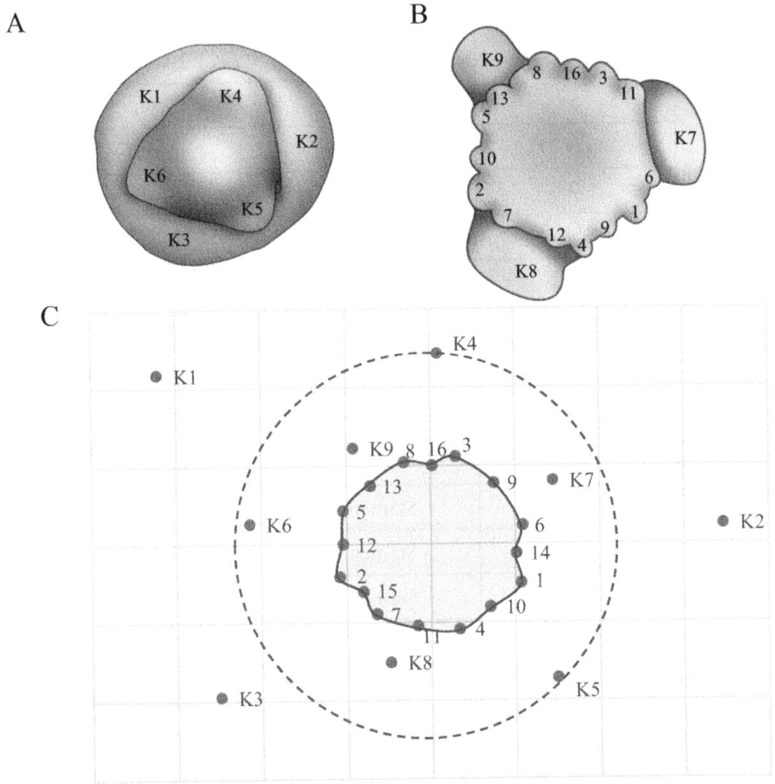

Figure 6.10. *(A, B) Initiation of the three tepal "whorls" and the first ring of stamens in Magnolia denudata, from SEM micrographs (redrawn from Erbar and Leins 1994). Notice the decrease in size of tepals. Tepals (K) and stamens are numbered according to their supposed location on the generative spiral according to Erbar and Leins. (C) Model of the initiation of the perianth and first stamens in M. denudata. The dashed circle shows the second tepal "whorl". Tepals (K) and stamens are numbered in order of appearance.* $r_0 = 0.35$, $\rho = 0.25$ and $r_{asympt} = 1.02$

Primordium number	Organ	Plastochron ratio
5	K1	1.12
6	K2	1.10
7	K3	1.30
8	K4	1.07
9	K5	1.06
10	K6	1.30
11	K7	1.04
12	K8	1.04
13	K9	1.30
14	Stamen	1.011
–	Stamen	–

Table 6.3. *Plastochron ratios used to model floral organ initiation in Magnolia denudata*

To model the initiation of floral organs in *M. denudata*, we used the plastochron ratios corresponding to a permuted phyllotaxis (Figure 6.2), while increasing the plastochron ratio between "whorls" (r = 1.30; Table 6.3). For the initial conditions of the model, the first tepal was considered to be the fifth organ of the ontogenetic spiral in continuity with the vegetative spiral.

The perianth consists of three trimerous "*pseudo-whorls*"; within each of them, the divergence angles are close to the golden angle (Figure 6.10). The spiral origin of the "whorls" of *M. denudata* is beyond doubt, and the structures thus formed cannot in any way be confused with true whorls like those of *Anaxagorea crassipetala* (Annonaceae; Figure 6.14). This is why we call them pseudo-whorls.

In Figure 6.10(B), stamen primordia are numbered according to their supposed position on the generative spiral, corresponding to the golden divergence angle (*as per* Erbar and Leins (1994), irrespective of their real temporal sequence). Erbar and Leins noticed that stamen primordia 1, 2, 3, 4 and 5 matched the temporal sequence. After that there is a break: primordium 8 of the generative spiral initiates before primordia 6 and 7. According to Erbar and Leins (1994), the possible causes for this break may be the large gaps between the three perianth members of the last "whorl" and the considerable size difference between tepal and stamen primordia.

The plastochron ratio separating successive stamens determines the number of organs within the first ring: with r = 1.011, this number is approximately 16 (Figure 6.10(C)), matching SEM data (Figure 6.10(B)). A higher plastochron ratio would lead to larger stamens, which would position some of the first 16 stamen primordia on an inner ring.

From the centric representation, we can see that the angular arrangement of the stamens is regular until the eighth stamen (by chronological order) (Figure 6.10(C)). Then, our simulation shows permutations of the stamen numbers on the first ring of the androecium, as observed by Erbar and Leins (1994). The location and the permutations of the stamens are very sensitive to parameter values. The geometric primordia sequence differs from the chronological sequence: the divergence angles between primordia are very unequal and do not converge toward a noble angle (not shown).

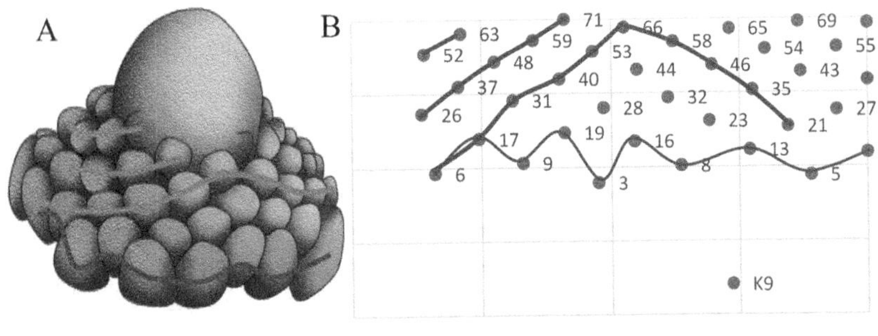

Figure 6.11. *(A) Initiation of stamens in Magnolia denudata (redrawn from Erbar and Leins 1994) with a pseudo-whorl (in red) and pseudo-parastichies (in blue). (B) Cylindrical representation according to model of stamen arrangement in M. denudata; the first pseudo-whorl of stamens is in red and the same pseudo-parastichies as in A are in blue*

The SEM micrograph of the androecium of *M. denudata* reveals irregular parastichies and quasi-symmetric phyllotaxis: the angles of right and left spirals are approximately equal (Figures 6.11(A) and (B)). The differences between primordia number along the pseudo-parastichies are not equal unlike those in Fibonacci spirals: for instance, from primordium 6: +11; +14; +9; +13; +13 (Figure 6.11(B)). This example illustrates the effect of discontinuity in plastochron ratios on phyllotaxis, that is, having large plastochron ratios between trimerous perianth pseudo-whorls and very small plastochron ratios between numerous small stamens.

An unusual arrangement of the gynoecium has been described in *Magnolia acuminata* (Zagórska-Marek 1994), illustrating the extraordinary diversity of phyllotaxis within the genus Magnolia. The divergence angle was calculated to be 132.4°. This angle is one of those minimizing the inhibition potential under the constraint of decreasing plastochron ratios (Figure 5.1) as well as packing efficiency (Figure 3.4). Phyllotaxis is in a (8, 11) mode. On the phyllotaxic frieze (Figure 4.4), we find that this mode optimizes primordia arrangement for a plastochron ratio ranging from 1.043 to 1.019.

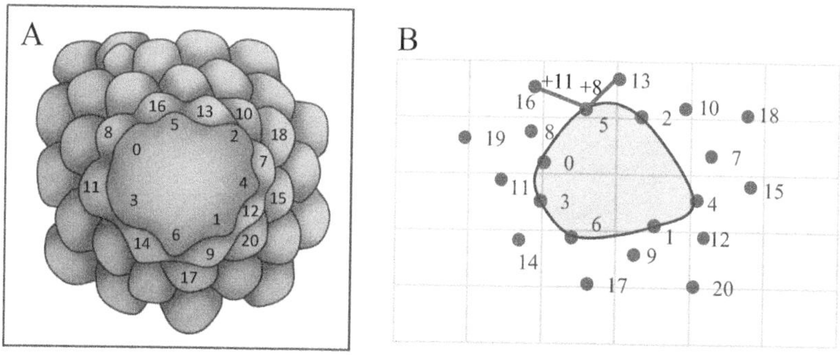

Figure 6.12. *(A) The gynoecium meristem of Magnolia acuminata, redrawn from Zagórska-Marek (1994) and (B) model of carpel initiation in M. acuminata (r = 1.041)*

For the simulation, we decided to space 20 stamen primordia at an angle of 132°, the data concerning the predecessors of those numbered by Zagórska-Marek being unknown, and apply our model with a constant plastochron ratio (r = 1.041). The phyllotaxis mode for carpels is in fact (8, 11): 5 + 8 = 13; 5 + 11 = 16 (Figure 6.11).

According to the van Iterson diagram (Douady and Couder 1996a) and our potential map (Figure 5.1), this mode cannot be reached continuously by following the decrease of the plastochron ratio from a distichous origin (see quasi-bifurcations, Chapter 5). The branch leading to it is disconnected from the (3, 5) line. Nevertheless, it is possible to observe this phyllotaxis mode in real life.

6.4.1.3. *Liriodendron*

Liriodendron tulipifera, also known as the tulip tree, is a large tree (up to 60 m tall) native to eastern North America whose flower shape resembles that of a tulip.

Flowers of *L. tulipifera* have nine tepals arranged in three pseudo-whorls: according to Erbar and Leins (1994), the first three plastochrons, which are short, are followed by a longer plastochron, followed by three short plastochrons, a longer one, and then again three short plastochrons. Each "whorl" of the trimerous perianth shows clear indications of a spiral initiation sequence.

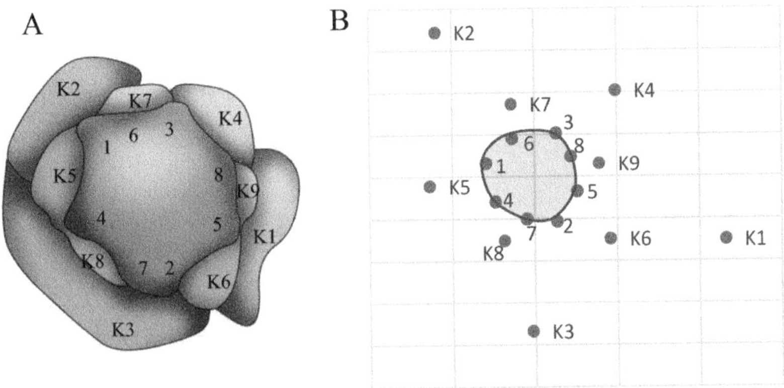

Figure 6.13. *(A) The floral meristem of Liriodendron tulipifera redrawn from an SEM micrograph from Erbar and Leins (1994) and (B) simulation of floral organ initiation in L. tulipifera, with r_0 = 0.35, ρ = 0.18 and r_{asympt} = 1.02. Tepals (K) and stamens are numbered by order of initiation*

Note that all stamens (from primordium 14) have the same plastochron ratio.

To model the initiation of floral organs in *L. tulipifera*, we used the plastochron ratios corresponding to Fibonacci spiral phyllotaxis and adding a long plastochron ratio (r = 1.30) between "whorls". The simulation reproduced the organ initiation sequence described by Erbar and Leins (1994; Figure 6.13). For the initial conditions of the model, the first sepal was considered to be the fifth organ of the generative spiral in continuity with the vegetative spiral (Table 6.4).

Primordium number	Organ	Plastochron ratio
5	K1	1.16
6	K2	1.14
7	K3	1.30
8	K4	1.10
9	K5	1.09
10	K6	1.30
11	K7	1.07
12	K8	1.06
13	K9	1.30
14	Stamen	1.03
–	–	–

Table 6.4. *Plastochron ratios used to model floral organ initiation in Liriodendron tulipifera*

The perianth thus consists of three trimerous pseudo-whorls; within each pseudo-whorl, the divergence angles are close to the golden angle (Figures 6.13(A) and (B)). The spiral origin of the "whorls" of *L. tulipifera* is beyond doubt, and the structures thus formed cannot in any way be confused with true whorls like those of *Anaxagorea crassipetala* (Annonaceae; Figure 6.14).

6.4.2. *Annonaceae*

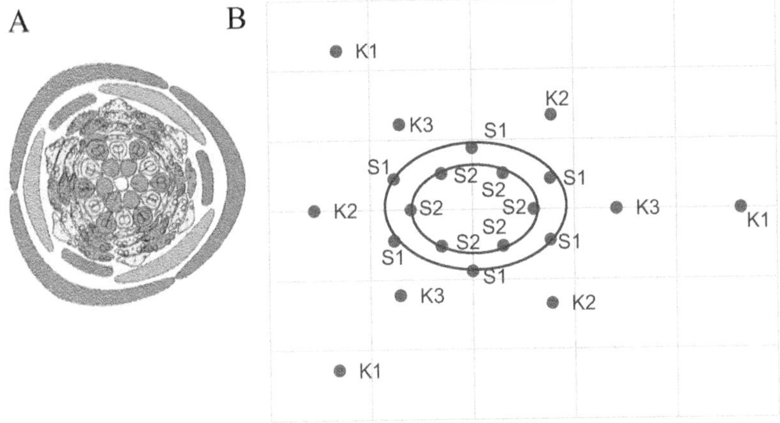

Figure 6.14. *(A) Drawing of a microtome section of the androecium and gynoecium of Anaxagorea crassipetala just before anthesis (from Endress and Armstrong 2011). Tepals have been added. Blue and green: tepals; yellow and red: stamens and carpels orthostichies. (B) Model of floral organ initiation in A. crassipetala, up to the second stamen whorl. Organs (K = tepals, S = stamens) are numbered according to their corresponding whorl*

Annonaceae is the largest family in the order Magnoliales with 100 genera and ~2,500 species. They are trees, shrubs and lianas from tropical and subtropical areas.

The flower of *Anaxagorea crassipetala* is composed of nine tepals, 42 stamens (of which 6–12 are sterile, i.e. staminodes) and 19 carpels. The three tepal whorls are trimerous. The first androecium whorl is composed of six stamens in "double position" alternating with the three inner tepals. The androecium consists of seven whorls of six stamens and the gynoecium of three whorls of six carpels. The androecium and the gynoecium form 12 regular orthostichies. Sometimes an additional carpel forms at the center of the floral apex (Endress and Armstrong 2011).

The perianth organs are initiated with very short plastochrons, but nevertheless conspicuous. By the time the first stamens are initiated, the three inner tepals are still much smaller than the six outer tepals.

Primordium number	Plastochron ratio	Primordium number	Plastochron ratio
K1	1	S1	1
K1	1	S1	1
K1	1.7	S1	1.4
K2	1	S2	1
K2	1	S2	1
K2	1.1	S2	1
K3	1	S2	1
K3	1	S2	1
K3	1.2	S2	1.2
S1	1	S3	1
S1	1	–	–

Table 6.5. *Plastochron ratios used to model floral organ initiation in Anaxagorea crassipetala*

By slitting diagonally the apex of *Epilobium hirsutum* (Onagraceae) with normally decussate phyllotaxis, Snow and Snow (1935) found that a majority of new apices developed spiral phyllotaxis. Richards (1951) showed that this change occurred even though there had been no change in the plastochron ratios. Thus, a change in phyllotaxis from decussate to spiral may appear just if the symmetry of the initial conditions changes.

Box 6.2. *The Snow and Snow (1935) experiment*

The process of initiating a whorl is very different from the process generating a spiral: the initial conditions (first whorl) have to be perfectly symmetrical and all primordia of this whorl must be initiated simultaneously. Thus, when the first primordium of the second whorl is initiated, the inhibition potential exhibits three minimums with equal inhibition potentials due to the perfect symmetry of the first whorl. Therefore, three primordia will initiate simultaneously at these minimums (Figure 6.15(A)).

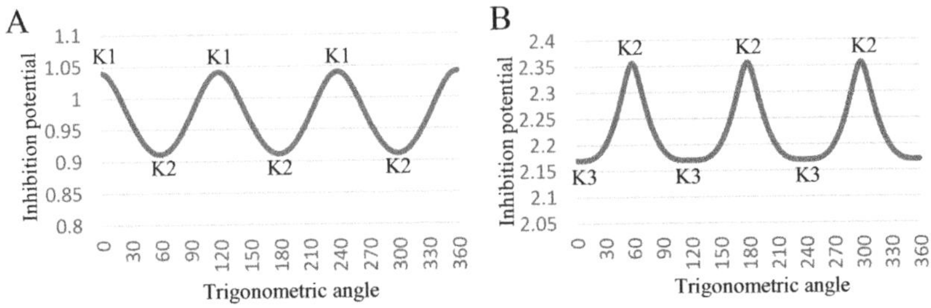

Figure 6.15. *Anaxagorea crassipetala. The inhibition potential profile on the meristem front at the initiation of (A) the second tepal whorl: the angular locations of existing maxima are given at K1; tepals K2 will be located at the minimums of the inhibition potential, and of (B) the third tepal whorl: the angular locations of existing maxima are given at K2; tepals K3 will be located at the minimums of the inhibition potential*

> According to Staedler and Endress (2009), the "double position" of organs in certain whorls is a striking characteristic: "two or more organs are in a position where only one is expected". It most often appears during the transition between the perianth and the androecium, but also sometimes within the perianth or the androecium, or even at the onset of gynoecium initiation. This phenomenon might be explained by the fact that doubled organs are smaller than those of the preceding whorl. It is a sign of great flexibility: indeed, this "dédoublement" can take place at different stages of floral development or in one whorl but not in the next.

Box 6.3. *Dédoublement (de Candolle 1813; Staedler and Endress 2009)*

In *A. crassipetala,* dédoublement occurs only in the fourth whorl (S1). This is due to the small plastochron ratio between whorls K2 and K3 (r = 1.1; Table 6.5), so that the inhibitory field generated by K2 is still strongly felt at the initiation of S1. Stamens in S1 are therefore doubled because of this double influence of K2 and K3 (Figure 6.16).

By contrast, since the plastochron ratio between the first (K1) and the second tepal whorl (K2) is high (r = 1.7), the inhibition field of the first whorl has no effect on the initiation of organs in the third whorl (K3), and so they are not doubled (Figure 6.15(B)).

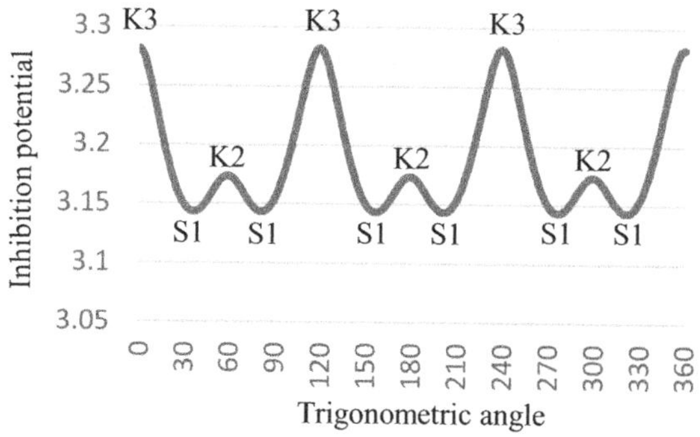

Figure 6.16. *Profile on the meristem front of the inhibition potential at the initiation of fourth whorl producing a dédoublement of stamens in Anaxagorea crassipetala*

Annona hexapetala (synonym *Artabotrys hexapetalus*) is a climbing shrub native to tropical Asia (southern India, Sri Lanka), whose highly fragrant flowers are used to flavor tea.

In *A. hexapetela*, the perianth is composed of two trimerous whorls and the first six stamens form simultaneously exactly at the corners of the hexagonal floral apex (Figure 6.17(A)). Subsequent stamens are initiated on a spiral that presents more or less regular parastichies (Figure 6.18(A); Erbar and Leins 1994).

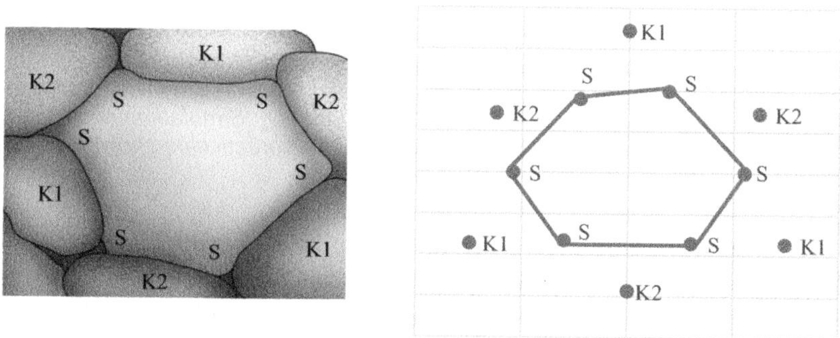

Figure 6.17. *(A) The floral meristem of Annona hexapetala where the two tepal whorls (K1 and K2) and the first six stamens are visible, redrawn from Erbar and Leins (1994) and (B) model of floral organ initiation*

This arrangement is similar to that of *Cabomba furcata* (Nymphaeales) (Walch and Blaise 2022a), but whereas *C. furcata* has only six stamens and three carpels, *A. hexapetala* produces a large number of stamens and carpels.

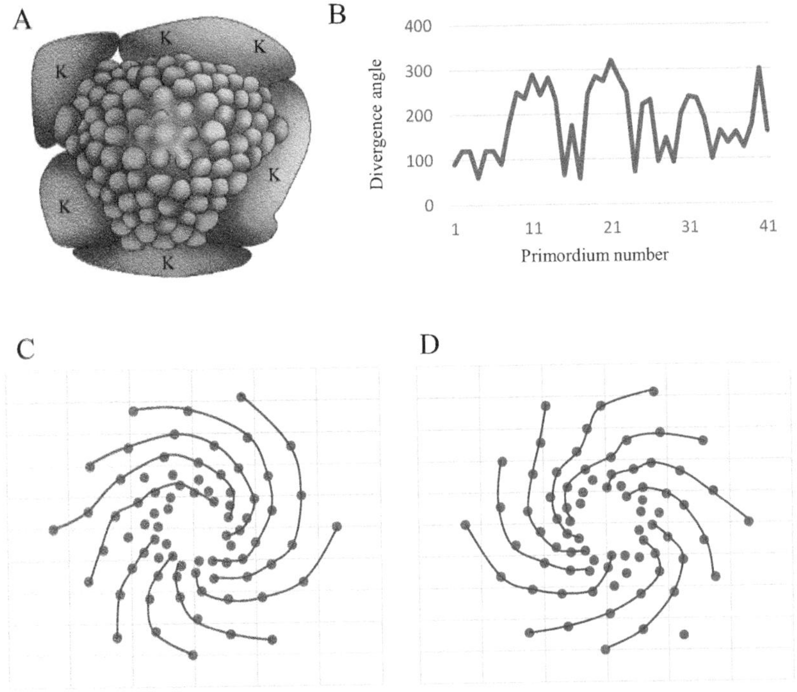

Figure 6.18. *(A) Sepal and stamen primordia in Annona hexapetala, redrawn from a SEM micrograph from Erbar and Leins (1994). (B) The profile of divergence angle up to primordium 41. (C and D) Simulation of stamen initiation with anti-clockwise and clockwise pseudo-parastichies shown*

To model the initiation of floral organs in *A. hexapetala*, we used the plastochron ratio for two pure trimerous whorls of tepal, then a unique plastochron ratio for very small stamens (Table 6.6). This was sufficient to produce a quasi-symmetric androecium with an equal number of right and left pseudo-parastichies (Figures 6.18(C) and (D)).

Primordium number	Plastochron ratio
K1	1
K1	1
K1	1.2
K2	1
K2	1
K2	1.3
S	1.02
–	–

Table 6.6. *Plastochron ratios used to model floral organ initiation in Annona hexapetala*

6.5. Laurales

The Laurales comprise approximately 2,500 species belonging to 85–90 genera grouped in seven families. Most species are tropical and subtropical trees or shrubs, although a few genera reach the temperate zone. The best-known species in this order belong to the family Lauraceae (e.g. bay laurel, cinnamon, avocado and sassafras) plus the ornamental shrub *Calycanthus* from the Calycanthaceae.

Members of the Magnoliales and the Laurales present "series" (Staedler et al. 2007), that is, sets of organs of the same type that look like whorls but with a less regular arrangement because neighboring organs are not equidistant. The number of organs in a series belongs to the Fibonacci sequence, but unlike in whorled phyllotaxis, the series is organized along more than one turn: approximately 1.5 turn for five organs, approximately two and a half turns for eight organs, etc. The divergence angle between an organ and its successor on the generative spiral is approximately 137° (Staedler et al. 2007). The question is: how can the Fibonacci sequence appear in these pseudo-whorled phyllotaxis? *We believe that these "series" correspond to meristem fronts developing according to a Fibonacci phyllotaxis* (Figure 6.6).

6.5.1. *Atherospermataceae*

Members of the Atherospermataceae, commonly known as southern sassafras, are trees or shrubs from temperate to tropical regions of the southern hemisphere. They are found in Australia, New Guinea, New Caledonia, New Zealand and Chile. The wood of humid forest species is harvested for construction and for manufacturing refined cabinets.

Daphnandra (Atherospermataceae) is endemic to Queensland and New South Wales in Australia. In *Daphnandra repandula the first two tepals are opposite*; subsequent floral organs show Fibonacci phyllotaxis. *The perianth and androecium form series of eight organs* (16 tepals, eight stamens, seven staminodes and 11 carpels), with the average divergence angle corresponding to the Fibonacci pattern: 137.6° ± 8.5° (Staedler and Endress 2009).

We made the hypothesis that *Magnolia* and *Liriodendron* take advantage of the continuity between the vegetative and the reproductive spiral. This is not the case in Atherospermataceae since, in this family, stem phyllotaxis is decussate. With the canonical parameters starting with two opposite organs, a large number of organs would be needed before obtaining a series of eight organs: indeed, this requires a plastochron ratio lower than 1.06 (Figure 6.6). The meristem would need to be very big. In actual fact, the meristem of *D. repandula* is very tight, with lower plastochron ratios as we can observe on Figure 6.19A. This arrangement is obtained with *a smaller front radius*: $R_0 = 0.5$ (Table 6.7; Figure 6.19C).

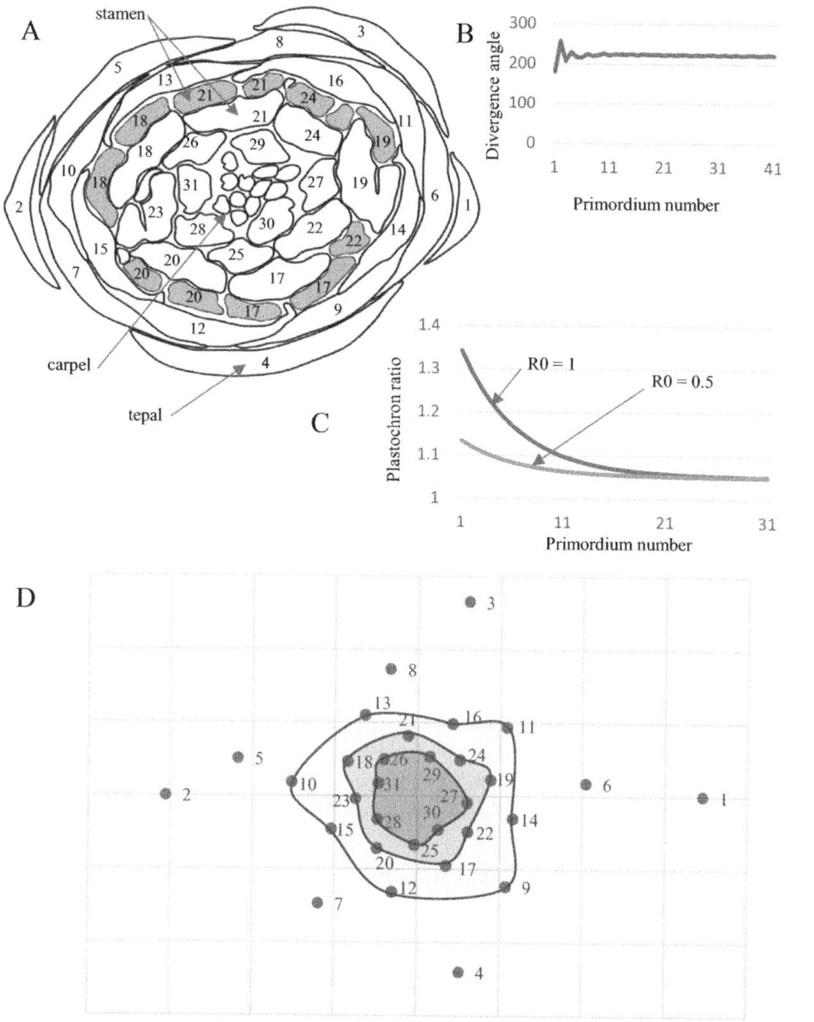

Figure 6.19. *(A) Transverse section of the flower bud of Daphnandra repandula, from Staedler and Endress (2009). (B) Evolution of the divergence angle and (C) plastochron ratio versus primordium number. (D) Model of D. repandula showing organs belonging to distinct series. R_0 = 0.5, r_0 = 0.18, ρ = 0.1 and r_{asympt} = 1.05*

Primordium number	Plastochron ratio	Primordium number	Plastochron ratio
1	1.13	13	1.06
2	1.12	14	1.06
3	1.11	15	1.06
4	1.10	16	1.06
5	1.09	17	1.05
6	1.08	18	1.05
7	1.08	19	1.05
8	1.07	20	1.05
9	1.07	21	1.05
10	1.07	22	1.05
11	1.06	–	–
12	1.06		

Table 6.7. *Plastochron ratios used to model floral organ initiation in Daphnandra repandula*

According to Staedler et al. (2007), stamens appear in series of five to eight in *Chimonanthus*, in series of 13 in *Calycanthus floridus*, *Sinocalycanthus* and *Idiospermum*, and in series of 21 in *Calycanthus occidentalis*. According to the phyllotaxic frieze, the rising phyllotaxis of the different species is the consequence of smaller plastochron ratios (see Chapter 4), which are only possible with small front radius.

Daphnandra micrantha is composed of six decussate tepals followed by a whorl of doubled tepals, and four stamens, eight staminodes and four carpels in alternating tetramerous whorls (Staedler and Endress 2009).

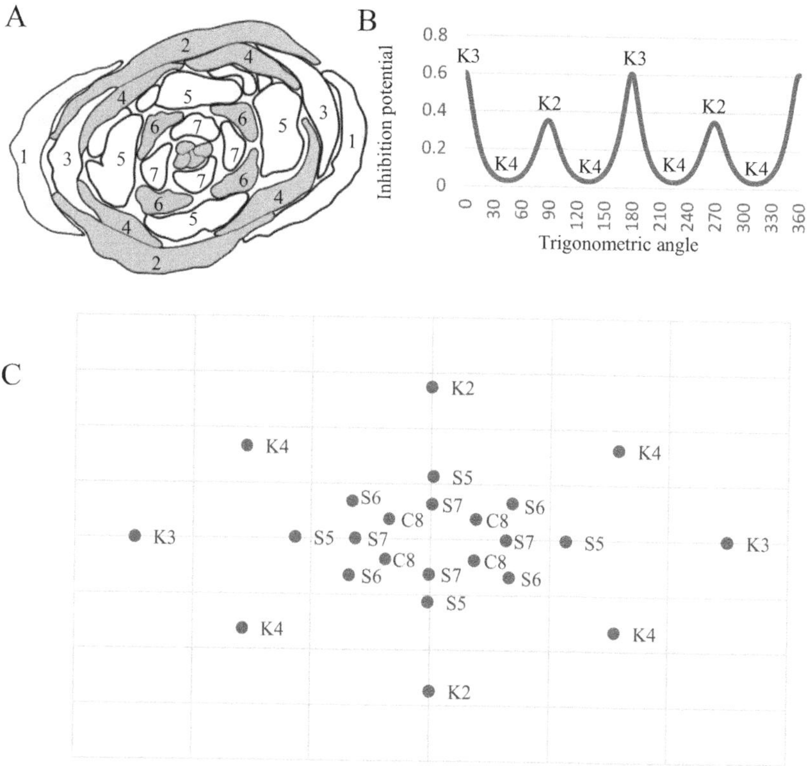

Figure 6.20. *(A) Transverse section of a flower bud of Daphnandra micrantha, from Staedler and Endress (2009). Floral organ numbers correspond to individual whorls. (B) Inhibition potential profile on the meristem front at the dédoublement in the fourth tepal whorl. Peaks represent the maxima of the inhibition potential produced by existing primordia (K2 and K3). K4 primordia are located at the minimums of the inhibition potential. (C) Model of floral organ arrangement*

To model the initiation of floral organs in *D. micrantha*, we used the plastochron ratios corresponding to perfect whorled phyllotaxis (Table 6.8).

Primordium number	Plastochron ratio	Primordium number	Plastochron ratio
K2	1	S6	1
K2	1.1	S6	1
K3	1	S6	1
K3	1.1	S6	1.5
K4	1	S7	1
K4	1	S7	1
K4	1	S7	1
K4	2	S7	1.2
S5	1	C8	1
S5	1	C8	1
S5	1	C8	1.05
S5	1.2		-

Table 6.8. *Plastochron ratios used to model floral organ initiation in Daphnandra micrantha*

The first tepal whorls are decussate. The plastochron ratio at the transition from K2 to K3 and K3 to K4 is low enough for the influence of K2 tepals to still be felt during the initiation of K4, resulting in a dédoublement of tepals in this whorl. Subsequent whorls are alternate tetramerous whorls (Staedler and Endress 2009) (Figure 6.20).

Doryphora (Atherospermataceae) includes four endemic species from Australia and New Caledonia. The flower of *Doryphora aromatica* has four tepals (numbered 1, 1, 2 and 3 by Staedler and Endress (2009)), five stamens (4-8), nine staminodes (9-16, 19) (five of which have glands and four do not) and eight carpels. The tepals arise as two decussate pairs while the other organs appear to be arranged in a Fibonacci spiral with an average divergence angle of 139.7° ± 36°. The androecium forms *two series of five organs* (five stamens and five glandular staminodes). "Unexpectedly", according to Staedler and Endress (2009), on the generative spiral the first two carpels (primordia 17 and 18) are *followed* by the last staminode (primordium 19), since the carpels can be recognized as being smaller than the staminodes (Figure 6.21A).

D. aromatica is modeled with a decrease in plastochron ratio that is faster than in the canonical sequence so that it generates an *M-shaped* divergence angle profile (Figure 6.21(B)).

Primordium number	Plastochron ratio	Primordium number	Plastochron ratio
K1	1.31	S10	1.03
K1	1.25	S11	1.03
K2	1.21	S12	1.03
K3	1.18	S13	1.03
S4	1.15	S14	1.02
S5	1.13	S15	1.02
S6	1.11	S16	1.02
S7	1.09	S17	1.01
S8	1.08	C18	1.01
S9	1.07	C	…

Table 6.9. *Plastochron ratios used to model floral organ initiation in Doryphora aromatica*

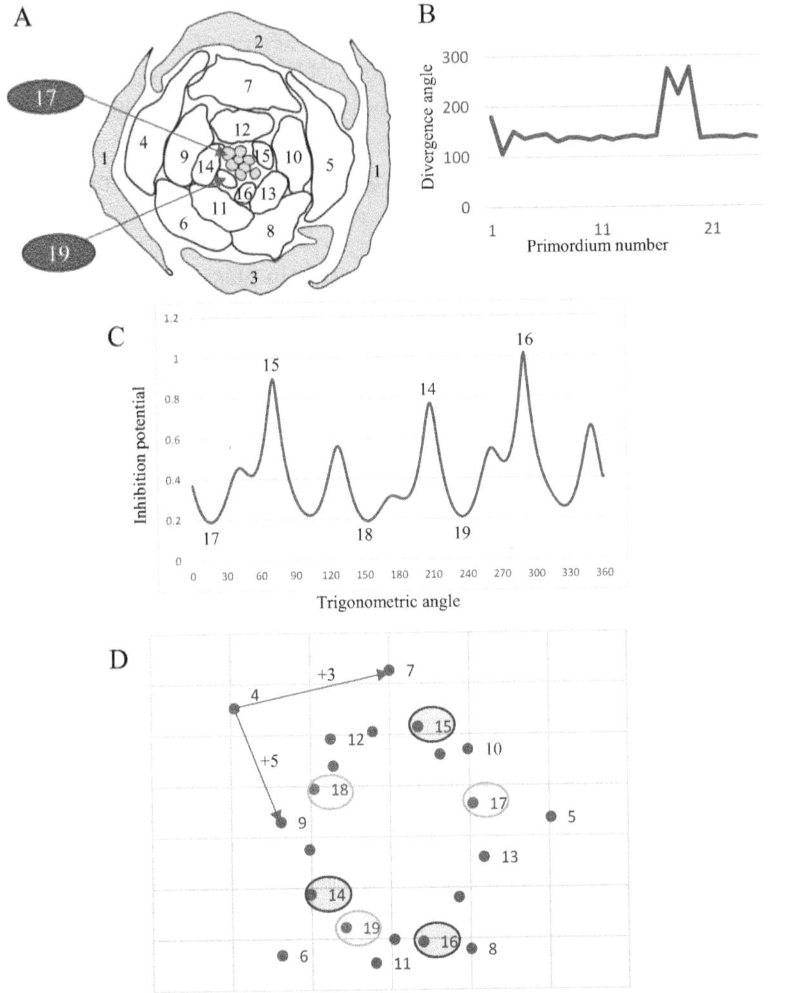

Figure 6.21. *(A) Transverse section of a flower bud of Doryphora aromatica (redrawn from Staedler and Endress 2009). Floral organs are numbered according to the Fibonacci sequence irrespective of their temporal sequence. (B) M-shaped divergence angle profile according to the model. (C) Inhibition potential profile on the meristem front at initiation of primordium 17. (D) Centric representation of organ arrangement.* r_0 = *0.35*, ρ = *0.30 and* r_{asympt} = *1.02*

To model the initiation of floral organs in *D. aromatica*, we used the plastochron ratio corresponding to a permuted phyllotaxis starting with two opposite tepals (Table 6.9; $r_0 = 0.35$, $\rho = 0.30$ and $r_{asympt} = 1.02$). The M-shape of the divergence angle profile corresponds to primordia permutation. At the initiation of the last stamen (primordium 17), three locations have approximatively the same inhibition potential, corresponding to the primordia 17, 18 and 19 (Figure 6.21C). Thus, the last stamen (S17) can be located where primordium 19 should be according to a Fibonacci sequence (Figure 6.21(D)).

6.5.2. *Monimiaceae*

Members of the Monimiaceae are shrubs, small trees or rarely lianas of the tropics and subtropics, mostly of the southern hemisphere. Monimiaceae is the most diverse family of Laurales, with flowers ranging from bisexual, spiral flowers with highly developed tepals to highly specialized unisexual dimerous flowers. Flowers are spiral (*Peumus*) or decussate (*Tambourissa, Kibara*).

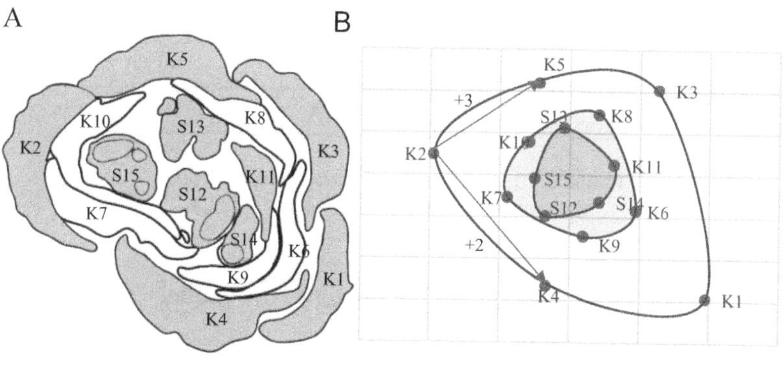

Figure 6.22. *(A) Transverse section of the flower bud of a male flower of Peumus boldus (redrawn from Staedler and Endress 2009) and (B) model of floral organ arrangement. Not all stamens are developed at this stage. K11 is part of the first series of stamens.* $r_0 = 0.35$, $\rho = 0.18$ *and* $r_{asympt} = 1.04$

Peumus boldus (boldo) is native to Chile. Its leaves, which have a strong woody aroma, are used for culinary purposes mainly in Latin America.

Male flowers are composed of 11 tepals and 36 stamens with an average divergence angle close to 137° ± 11° "in accordance with the Fibonacci spiral arrangement" (Staedler and Endress 2009). *Floral organs form series of five organs*: two series of five tepals, one series that includes the last tepal (K11) and four stamens, then seven series of stamens (Staedler and Endress 2009).

Primordium number	Plastochron ratio	Primordium number	Plastochron ratio
B	1.33	K8	1.10
B	1.28	K9	1.09
K1	1.24	K10	1.08
K2	1.21	K11	1.07
K3	1.18	S	1.07
K4	1.16	S	1.06
K5	1.14	S	1.06
K6	1.12	S	1.06
K7	1.11	S	1.05

Table 6.10. *Plastochron ratios used to model floral organ initiation in Peumus boldus*

We used the canonical parameters to model floral organ arrangement in *P. boldus*.

According to Staedler and Endress (2009), in Laurales with spiral flowers the change from vegetative decussate to floral spiral phyllotaxis can depend on the location of flower in the inflorescence. In *Daphnandra repandula*, the first two tepals are the only one to be opposite before spiral arrangement (Figure 6.19(A)). In *Doryphora aromatica*, the transition from decussate to spiral phyllotaxis occurs at the transition from the perianth to the androecium (Figure 6.21(A)). In

Peumus boldus, the first tepals are already spiral (Figure 6.22(A)). That is why in the latter we initiated the model with two opposite bracts (Table 6.10). These plants thus change from decussate leaves to spiral flowers.

6.6. Discussion

Our models suggest that phyllotaxis in Magnoliales and Laurales should not be coded following the classic spiral/whorled dichotomy, but according to five types: regular spiral, permuted spiral, quasi-symmetric spiral, pseudo-whorl and whorl.

These types are defined as follows:

– *Regular spiral* (e.g. *Magnolia soulangeana*): the sequence of plastochron ratios is the canonical sequence. There is an unequal number of right and left perfect parastichies (e.g. 5 and 8). The sequence of primordia on the generative spiral corresponds to their chronological order.

– *Permuted spiral* (e.g. *Doryphora aromatica*): there is an unequal number of right and left pseudo-parastichies. The primordia sequence on the generative spiral is different from the chronological order of initiation.

– *Quasi-symmetric spiral* (e.g. *Annona hexapetala*, *Nuphar pumila*; Walch and Blaise 2022a): there is approximatively equal number of right and left pseudo-parastichies (distorded parastichies).

– *Pseudo-whorl* (e.g. *Liriodendron*, *Daphnandra repandula*): the plastochron ratios of the spiral are low, which allows the constitution of whorls whose merism corresponds to terms of the Fibonacci sequence. There is a lack of symmetry.

– *Whorl* (e.g. *Daphnandra micrantha*): dimerous, trimerous, tetramerous whorls of organs in single or double positions. Primordia are initiated simultaneously within each whorl where the plastochron ratios are unitary. The symmetry is preserved, whether whorls are simple or double.

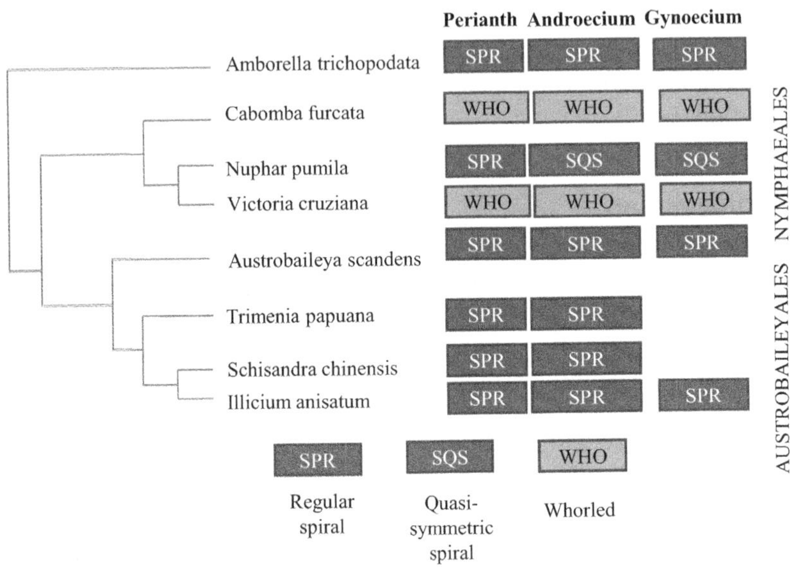

Figure 6.23. *Phyllotaxis in basal angiosperms from the ANA grade according to our model. Cabomba is trimerous and Victoria tetramerous (Walch and Blaise 2022a)*

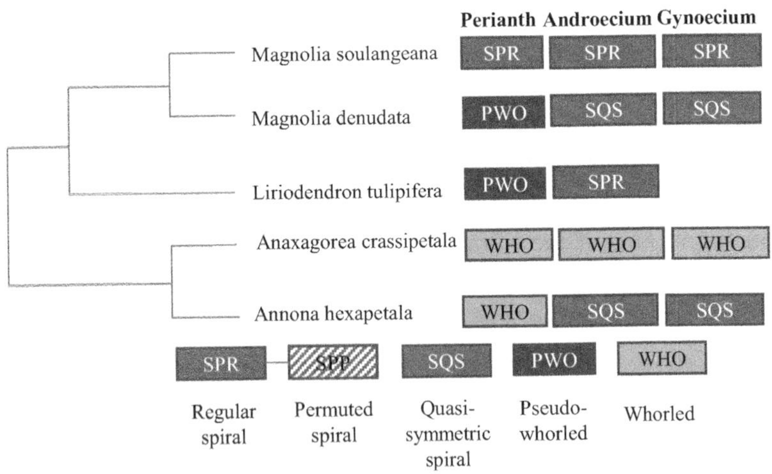

Figure 6.24. *Phyllotaxis in Magnoliales according to our model. Anaxagorea and Annona are trimerous*

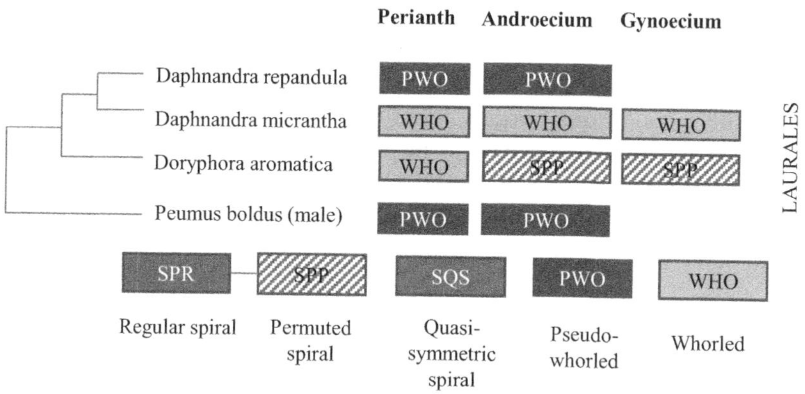

Figure 6.25. *Phyllotaxis in Laurales according to our model. Daphnandra micrantha is tetramerous*

6.6.1. *Regular spirals*

Fibonacci phyllotaxis requires a controlled decrease in plastochron ratios during meristem growth (the canonical sequence). It is mainly found in the ANA grade from *Amborella* to *Illicium* and in some Magnoliaceae such as *Magnolia soulangeana*.

6.6.2. *Permuted spirals*

When the transition between large tepals and the androecium is "abrupt", spiral phyllotaxis is "disrupted": stamens are not initiated in a regular spiral (Ronse de Craene 2022).

Staedler et al. (2007) recognized that in certain cases, it was difficult to interpret the ontogenetic spiral with a divergence angle supposedly very close to 137°. Indeed, the first carpel of *Doryphora aromatica* and *Hedycarya angustifolia* (Staedler and Endress 2009), but also of *Calycanthus occidentalis*, *C. floridus* and *Chimonanthus praecox* (Staedler et al. 2007), seems to be initiated before the last staminode or the last two staminodes, the other carpels being initiated after the last staminode. Despite this permutation, the position of organs corresponds to Fibonacci phyllotaxis. According to Staedler et al., the floral apex then reaches its maximum size and, under these "extreme" conditions, organ identity would depend on the local environment rather than on the position on the ontogenetic spiral if subsequent organs are small and not contiguous. Would this favor Plantefol's view (1949) that parastichies rather than the generative spiral determine organ initiation?

According to our models, if plastochron ratios decrease faster than in the canonical sequence, then the chronological sequence of the primordia no longer corresponds to their succession on the ontogenetic spiral with "golden" divergence angles. This is visible on a centric representation by the permutation of organs (Figure 6.2(C)). The divergence angles have a characteristic M-shaped profile (Figure 6.2(A)). This pattern is found in species with flowers where there is a significant size difference between tepals and stamens.

6.6.3. *Quasi-symmetric spirals*

In some species of basal angiosperms, the nature of floral phyllotaxis remains controversial. For instance, floral phyllotaxis in *Nuphar* has traditionally been thought to be spiral, a view supported by August Eichler (1878); however, Endress and Doyle (2015) have asserted that all Nymphaeaceae have flowers with complex-whorled (i.e. non-spiral) phyllotaxis. The criterion used by Endress is as follows: phyllotaxis is whorled when the inclination of the left and right parastichies is the same. According to the observations of El et al. (2020), the calyx of *Nuphar* is neither spiral nor composed of a single pentamerous whorl. As for the androecium, it is "fundamentally whorled, with more or less pronounced irregularities", the merism and number of whorls being unstable. An "ideal" condition for *Nuphar lutea* and *Nuphar pumila* would be an equal number of antepetal and alternipetal orthostichies. However, one flower of *Nuphar lutea* had 14 parastichies in one direction and 15 in the other, which would give it a "merism" of 14.5.

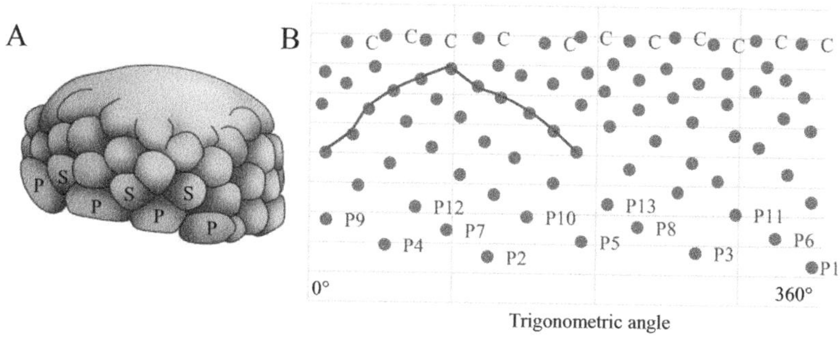

Figure 6.26. *(A) The floral meristem of Nuphar pumila (from El et al. 2020); (B) cylindrical representation of the modeled floral meristem of Nuphar pumila, showing quasi-symetric pseudo-parastichies (from Walch and Blaise 2022a). Px = petal; S= stamen; C = carpel*

According to our model, if the plastochron ratios decrease very rapidly, phyllotaxis becomes quasi-symmetrical. There is then a near equal number of left and right parastichies. This type of spiral phyllotaxis is different from the Fibonacci canon and can be mistaken for whorls. Indeed, the criterion used by Endress and Doyle is insufficient: an equal inclination of left and right parastichies differentiates whorled phyllotaxis from Fibonacci phyllotaxis, but not whorled phyllotaxis from certain types of irregular spirals.

6.6.4. *Whorls*

Whorled phyllotaxis requires initial symmetry and plastochron ratios equal to one within a whorl and strongly discontinuous between whorls. Whorled phyllotaxis is just a special case of the dynamical model that satisfies these constraints.

When a whorl is *symmetrical* (dimerous, trimerous or tetramerous), the inhibition potential it generates presents several absolute minimums having *equal* potential values at different angular positions. This is the case in the first whorl of *Anaxagorea crassipetala*. The next three primordia are *initiated simultaneously* at the three minimums such that nothing distinguishes one from the other from a physicochemical point of view (the inhibition potentials are identical). These primordia thus form a whorl, which in turn determine the initiation still symmetrical of primordia in a subsequent whorl (Figure 6.15).

6.6.5. *Dédoublement*

According to Staedler and Endress (2009), the "double position" of organs in certain whorls is the most striking characteristic of families belonging to the core Laurales. They suggest that this phenomenon is due to the fact that doubled organs are smaller than those of the preceding whorl. It is a sign of great flexibility: dédoublement can occur at different stages of floral development and in specific whorls (e.g. one whorl being double, the next simple).

Once again, according to our models, this phenomenon depends on plastochron ratios, for example, consider *Anaxagorea crassipetala*. Its dédoublement occurs in the fourth whorl (S1) and is the consequence of the plastochron ratio between consecutive whorls. In this case, the plastochron ratio between whorls 2 and 3 and between whorls 3 and 4 must be small (K2 to K3: r = 1.1 and K3 to S1: r = 1.2), so that the inhibitory field generated by K2 is still strongly felt at the initiation of S1. The double influence of whorls K2 and K3 produces double positions in the inhibition potential profile at the initiation of S1. The simplicity of this mechanism explains its flexibility.

6.6.6. *Pseudo-whorls*

Magnoliales and Laurales use a "strategy" to align organ identity with spiral phyllotaxis based on "series" (Staedler et al. 2007), that is, sets of organs of the same type that look like whorls but are less regular because they are derived from spiral initiation with golden divergence angles instead of rational angles dividing the circle into equal parts. Organs within one series are not initiated simultaneously but sequentially after a short plastochron, the initiation of the first organ of the next series being delayed by a longer plastochron. *Each cycle corresponds to a front of the spiral* which explains that the number of its organs is part of the Fibonacci sequence and could be improved by an adjustment stage.

The number of organs in the series is determined by the plastochron ratio, as specified in our phyllotaxic frieze. The androecium of *Doryphora aromatica* forms multiple series of five organs, eight organs in *Daphnandra repandula*, 13 organs in *Calycanthus floridus* and 21 organs in *Calycanthus occidentalis*.

We call theses "series" pseudo-whorls.

6.6.7. *Hemicycles*

Flowers that have certain organs arranged in a spiral and others in whorls are called hemicyclic. Magnoliales and Laurales show transitions from whorled to spiral phyllotaxis (e.g. *Annona hexapetala, Doryphora aromatica*) but the opposite is very rare. Sokoloff et al. (2018) proposed the terms "cyclospiral" for flowers with outer whorled and inner spiral phyllotaxis and "spirocyclic" to describe flowers with outer spiral and inner whorled phyllotaxis.

Spirocyclic flowers are very rare. True whorls suppose a very high level of symmetry, a very specific property. If this symmetry is broken by a period of spiral growth, it is difficult for the meristem to become whorled again. However, an example is found in eudicots in *Berberidopsis beckleri* (Berberidopsidaceae) where the perianth with Fibonacci phyllotaxis is followed by a perfect whorled androecium and gynoecium (Ronse de Craene 2017). Surprisingly, *Aquilegia vulgaris* (Ranunculaceae) is spirocyclic, a feature which seems not to have been noticed.

6.6.8. *Synorganization*

Angiosperm flowers are characterized by three types of organs that have the following basic functions: (1) protection and visual attraction (the perianth organs: tepals or sepals and petals), (2) male function (the androecial organs: stamens) and (3) female function (the gynoecial organs: carpels). These floral organs are modular structures, repetitive units of the same type. The number and arrangement of each type of organ is variable, although their sequence from the periphery to the center of the flower is highly conserved: perianth → androecium → gynoecium. This organization into a complex structure is called synorganization (Endress 2016).

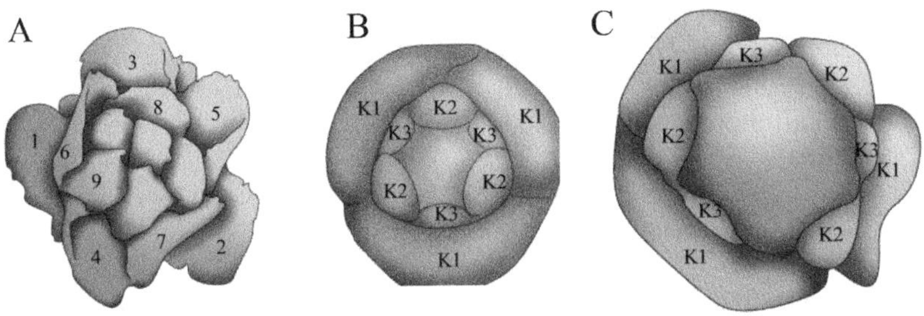

Figure 6.27. *Main tepal arrangement. (A) Spiral (Amborella trichopoda; according to Endress (2001)). (B) Whorled (Anaxagorea crassipetala from Endress and Armstrong (2011)) and (C) pseudo-whorled (Liriodendron tulipifera from Erbar and Leins (1994))*

In basal angiosperms, floral organs are arranged following two main types of phyllotaxis: spiral and whorled. Spiral phyllotaxis optimizes the packing of primordia (see Chapter 3). But there is no physical limit to the number of organs, so these numbers are labile: *Austrobaileyaceae* have 19–23 tepals, 7–11 stamens, 9–16 staminodes and 10–13 carpels; *Trimeniaceae* have up to 20 tepals, 7–16 stamens and a single carpel; *Illicium* has 7–33 tepals, 4–50 stamens and 5–21 carpels (Endress 2001).

Packing is less optimal in whorled phyllotaxis. However, it facilitates synorganization, organ identity being coherent with this type of phyllotaxis: the identity of the organs is the same throughout one whorl and therefore each species has a well-defined number of organs of each type.

Unlike most basal angiosperms, *Magnoliales and Laurales have enhanced synorganization inside their spiral flower*. The front of the meristem aligns more or less on the same circle a number of primordia, which is one of the terms of the Fibonacci series, making it possible to create pseudo-whorls whose organ numbers belong to the Fibonacci sequence (Figure 6.6). Thus, the perianths of certain Magnoliales (*Liriodendron*) and Laurales (*Daphnandra*) present the combined advantages of spiral and whorled phyllotaxis: optimization of packing and consistency between the position of the organs and their identity.

In Magnoliales, the spiral phyllotaxis of the vegetative shoot is transmitted by the spathe to the perianth. According to the phyllotaxis frieze, with relatively high plastochron ratios, it can produce trimerous pseudo-whorls. As it is well known, these flowers are large, with large tepals and numerous stamens and carpels.

In some Laurales (e.g. *Daphnandra repandula*), the diameter of the front is reduced, which allows a Fibonacci spiral with low plastochron ratios to appear soon after the last opposite organs (bracts or first tepals). Thus, a high phyllotaxis mode with pseudo-whorls of small organs is possible even in smaller flowers.

Eichler (1875) thus begins his introduction on the arrangement of floral parts: "The flower phyllomes are usually arranged in whorls (Quirl), rarely spirally. If they are throughout in whorls, the flower is cyclic; acyclic when the arrangement is throughout spiral, and hemicyclic when it is partly spiral and partly whorled".

His classification of flowers corresponds neither to the results of theoretical phyllotaxis nor to the mathematical definition of symmetries.

According to our mathematical model, there are two ways for a plant to implement the engine of phyllotaxis: either (1) each primordium is initiated after the previous one, without any symmetry constraint or (2) several primordia are initiated simultaneously under a strong symmetry constraint.

The latter corresponds to the initiation of the second whorl of the perianth of *Anaxagorea crassipetala* (Figure 6.28). According to the phyllotaxis model, primordia are initiated where the inhibition generated by pre-existing primordia is minimum. When the three primordia of the first whorl are initiated simultaneously at 120° from each other, the inhibition potential presents three simultaneous minimums at 120° from each other in strict alternation. Symmetry is transmitted from whorl to whorl. *Anaxagora crassipetala* is entirely whorled and polysymmetric, that is to say the flower has several axes of symmetry (in this case, three).

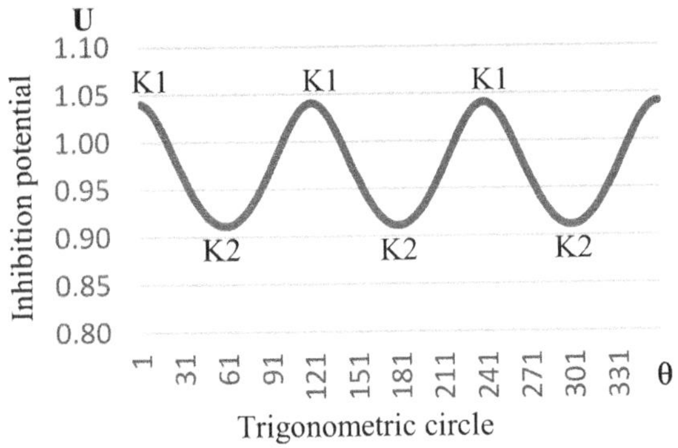

Figure 6.28. *Inhibition profile on the meristem front at the initiation of the second whorl in Anaxagorea crassipetala*

If one of the primordia of the second whorl was initiated before the other two, it would produce an inhibition field that would modify the field generated by the first whorl. The position of the next primordium (the fifth tepal) would be modified, and the ternary symmetry would be broken.

Spiral *Daphnandra repandula* looks a lot like a twin of the whorled *Daphnandra micrantha*. We know that the transient regime requires a small perturbation affecting the pair of opposite organs at the base of a spiral phyllotaxis. Does *D. micrantha* have flowers where this perturbation did not occur? These two species are the natural version of the Snow and Snow experiment. They show that the barrier between whorled and spiral phyllotaxis is not difficult to cross, from the former to the latter.

Thus, whorled phyllotaxis and the symmetry which corresponds to it do not tolerate from any perturbation.

By contrast, as we have seen in Magnoliales and Laurales, spiral phyllotaxis can evolve to become pseudo-whorled. This can be seen in the trimerous perianths of these early diverging angiosperm lineages, but also in the quincuncial perianths of many eudicots, where the order of initiation of the organs on the generative spiral can easily be recognized. Compared to purely spiral phyllotaxis, pseudo-whorled phyllotaxis can ensure a consistency between structure and function, between geometry and organ identity: this improves synorganization.

Pseudo-whorls are therefore not “lightly spiral whorls”, but “whorled spirals” and they are asymmetrical from a purely mathematical point of view.

The world of flowers is not divided between purely whorled and purely spiral, and whorled flowers are not the majority but minority. Nevertheless, it is clear that Eichler’s classification remains dominant to this day.

7

Inflorescences

7.1. Bracteole theory

In 1875, the German botanist A. W. Eichler studied at length the relationship between flowers and bracteoles ("Vorblätter. Anschluss und Einsatz der Blüthe"). When only one bracteole is present – in particular in monocots – the position of the first sepal/tepal (K1) is opposite the bracteole. When two bracteoles are present, he considered that the sepals were located on the same spiral as the bracteoles (α, the older bracteole, and β, the younger bracteole). He then described the different possible positions of the first sepals.

In the 20th century, Eichler's theory was mentioned, without any added value, in some university textbooks (e.g. Chadefaud and Emberger 1960; Guinochet 1965), as well as in Weberling's treatise on inflorescences (1989).

Eichler tried to explain sepal arrangement using Hofmeister's (1868) rule, that is, a new primordium is initiated where its predecessors leave enough space. This rule predicts an alternation of bracteoles and sepals. However, this is only one particular case of the general rule stating that primordia are initiated where the inhibition field is minimal, and the many exceptions forced Eichler to abandon the path opened up by Hofmeister.

For a color version of all the figures in this book, see www.iste.co.uk/walch/phyllotaxis.zip.

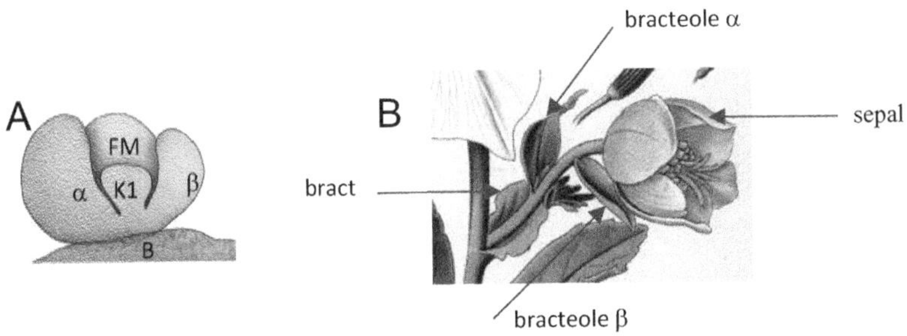

Figure 7.1. *(A) Initiation of the floral meristem (FM), bracteoles (α and β) and the first sepal (K1) in Aquilegia vulgaris (redrawn from Tucker and Hodges (2005)), with the older bracteole being larger. (B) Illustration of a flower of Helleborus niger (Thomé 1885). Bracteole α is under β on the pedicel*

If we hypothesize that bracteoles determine the arrangement of sepals, the dynamical model of phyllotaxis based on an inhibition field makes it possible to account for the numerous arrangements (trimery, tetramery, pentamery) found in monocots and eudicots. This hypothesis suggests that determining sepal arrangement is the first biological function of bracteoles, as intuited by Eichler in 1875. Our model of phyllotaxis explains these different cases in a "purely mechanical" way, thus responding to the challenge launched by Eichler 150 years ago (Walch and Blaise 2022b).

Nakagawa et al. (2020) modeled the effect of an adaxial-abaxial inhibition gradient on the position of the first floral organs. To do this, they created two fictitious inhibitory primordia (one adaxial and one abaxial), located at arbitrary distances from the center of the meristem. They reproduced different patterns of the arrangement of the outermost floral organs by varying the distance from the center of the meristem of the two fictitious primordia and the radius of the central region of the meristem (Figure 7.2). However, flowers with external inhibition sources that are located only on the dorso-ventral axis are rare, except in monocots corresponding to type 3B.

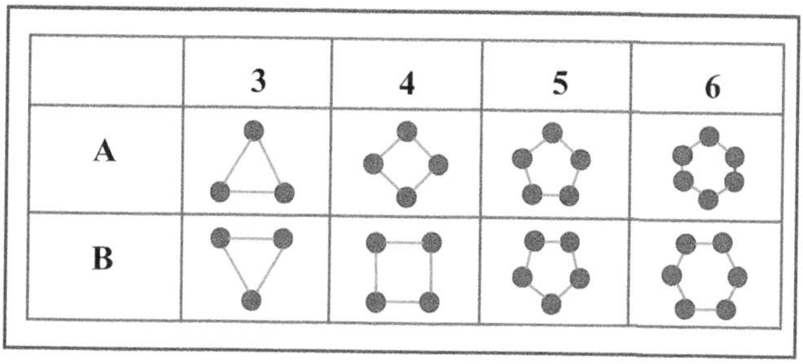

Figure 7.2. *Stylized types of arrangements of the outermost floral organs (sepals or outer tepals) (from Nakagawa et al. (2020)). 3–6: merosity; A: one adaxial organ; B: no adaxial organ*

We will use the typical arrangements defined by Nakagawa et al. (2020).

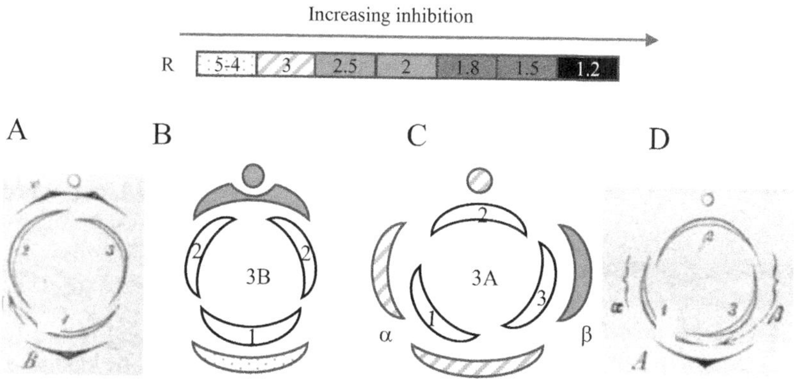

Figure 7.3. *(A) Diagram of a 3B arrangement in a monocot with a bikeeled adaxial bracteole (Eichler 1875). (B) Modeling of a flower with an abaxial bract and an adaxial bracteole (as in Iridaceae). The first floral organs have a 3B arrangement. Numbers correspond to the chronological order of initiation. The upper circle represents the axis of the flower, with the adaxial bracteole underneath, which exerts an influence on the position of future primordia. Effective distances from the center of the meristem (R) are coded from dotted (R = 4–5) to black (R = 1.2): the shorter the distance, the higher the inhibition intensity. As a consequence of the initial symmetry (see Chapter 6), the two dorsal sepals are initiated simultaneously. (C) Modeling of a 3A arrangement in a eudicot (Menispermum, Ranunculales). (D) Diagram of the 3A arrangement in eudicots (Eichler 1875)*

The distances used in the model should be understood as *effective distances* and not actual distances, as they are measured on the plant (see Chapter 4).

In monocots, the first sepal is initiated at 180° of the bracteole; this pattern can be considered to conform to Hofmeister's rule (Figure 7.3(A)). In trimerous dicotyledons, however, this rule does not apply, as the distances of bracteoles α and β (β being closer to the center as it is initiated after α) determine sepal arrangement (3A type) (other external organs are too distant to exert an influence). In general, the greater the influence of bracteole β, the closer the first sepal is to α, that is, the more the sepal position deviates from the median position between α and β, and therefore from a strict application of Hofmeister's rule (Figure 7.3(C)).

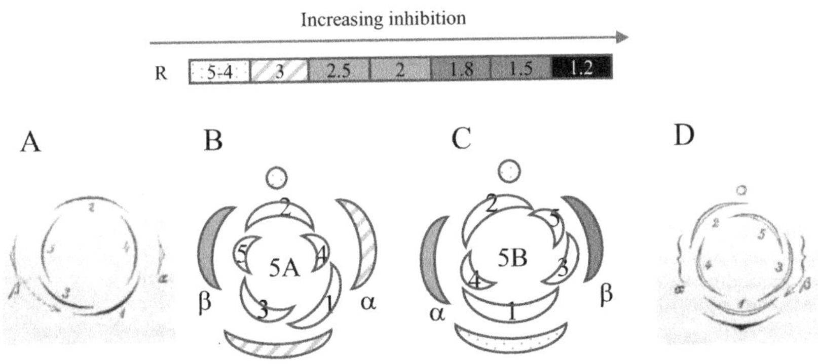

Figure 7.4. *(A) Diagram of a 5A arrangement (Eichler 1875). (B) Modeling of a 5A arrangement. (C) Modeling of a 5B arrangement. (D) Diagram of the 5B arrangement in Faboideae and Caesalpineae (Eichler 1878)*

In eudicots, if the bracteoles are located on the vegetative spiral (with a divergence angle close to 137°), sepals will be in a *"main" quincuncial arrangement (with K2 in an adaxial position*; Figure 7.4(A)), the most frequent arrangement in eudicots. The only sepal on the dorso-ventral axis is located on the adaxial side, corresponding to a 5A arrangement (Figure 7.2). K1 is furthest away from bracteole β (the second bracteole in order of initiation and the closest to the center of the floral meristem), since the strength of the inhibition of

β is greater than that of α (the first bracteole to be initiated and thus the furthest from the center of the floral meristem). If the strength of the inhibition of β is very high ($R_\beta = 1.5$; Figure 7.4(C)), then K1 is abaxial and the flower has a 5B arrangement, as is found in Fabaceae and Caesalpinioideae.

Eichler noted that organ arrangement was the same whether the bracteoles were permanent or ephemeral when he was able to observe the disappearance of one or both bracteoles. For instance, in *Baptisia australis*, the bracteoles disappear after all sepals have been initiated (Prenner 2004) (see Chapter 5).

Eichler also noted that different numbers of bracteoles can produce the same calyx arrangement, as seen in trimerous flowers, where a trimerous arrangement can be obtained when either one or two bracteoles are present, or even without a bracteole. More generally, several parametrizations of our model can produce the same sepal arrangement (*non-uniqueness* of the mathematical solution).

7.2. Inflorescences

Claßen-Bockhoff and Bull-Hereñu (2013) defined a conceptual framework for the organization of inflorescences based on an ontogenetic approach taking into account the different types of meristems and developmental pathways. Reproductive systems traditionally called inflorescences fall into two groups after vegetative development. First, inflorescences *sensu stricto*: we will take racemes as an example; second, floral units: we will take cymes as an example.

During vegetative development, the shoot apical meristem (SAM) gives rise to leaves and stem tissue. During the transition to flowering, the SAM is transformed into an *inflorescence meristem (IM)*, which in racemes generates at its periphery *floral meristems (FM)* according to a Fibonacci spiral arrangement (Figures 7.5(A) and (B); Rebocho et al. 2008).

The main inflorescence of *Antirrhinum majus* (Plantaginaceae) is a simple raceme (Figure 7.5(B)). *Racemose* systems can branch into lateral racemes to form compound racemes (Figure 7.5(C)).

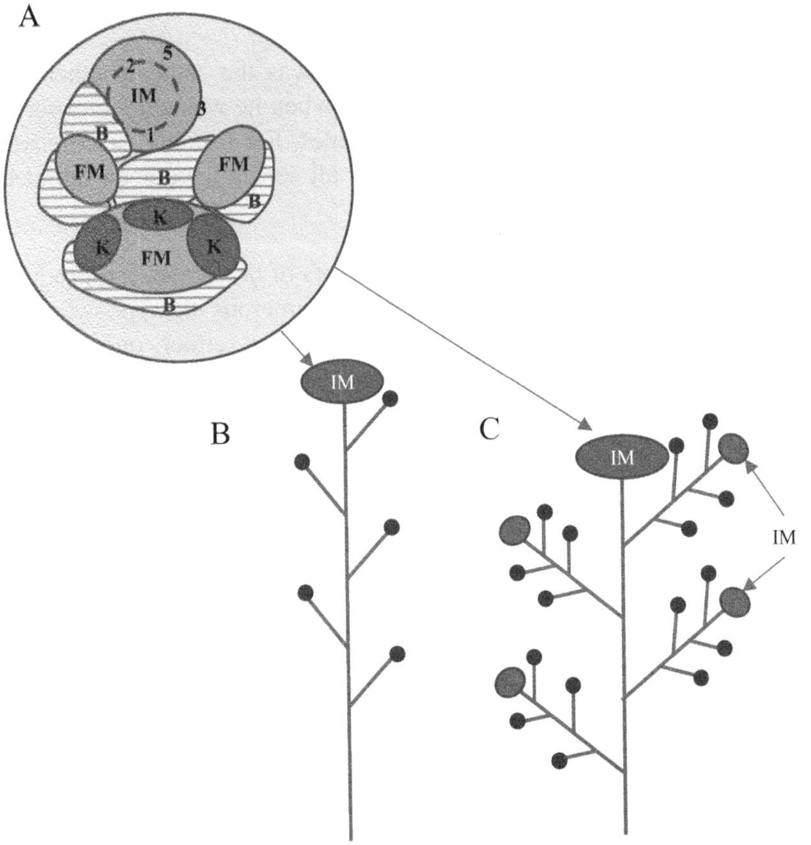

Figure 7.5. *(A) Development of a racemose inflorescence (see Figure 7.8A). (B) Simple raceme (e.g. Antirrhinum majus). (C) Raceme composed of lateral first-order racemes. IM = inflorescence meristem (or monopodial, in blue). FM = floral meristem. B = bract. K = sepal. Red flower = produced by an IM within a raceme*

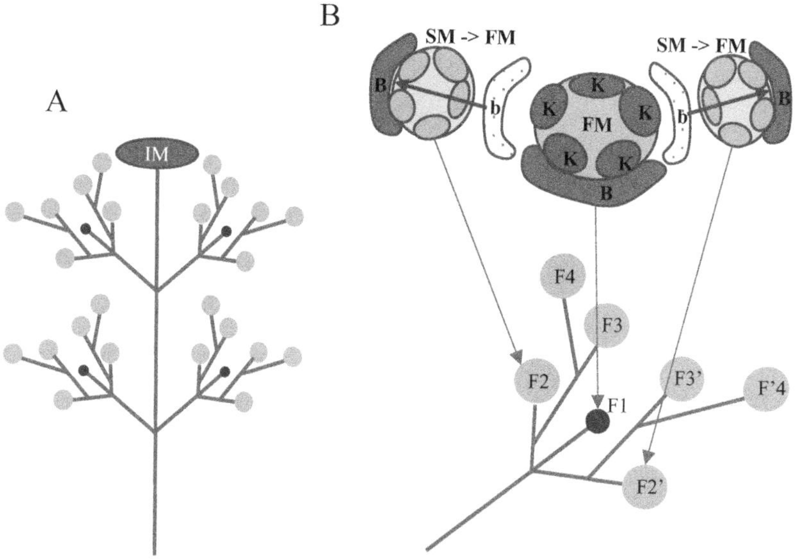

Figure 7.6. *(A) Thyrse (e.g. Salvia verticillata) = raceme composed of lateral cymes. (B) Cyme of a thyrse. IM = inflorescence meristem (or monopodial, in blue). SM = sympodial meristem. FM = floral meristem. B = bract. K = sepal. b = bracteole. Red flower = flower produced by an IM within a raceme. Orange flower = flower produced by a SM within a cyme. The cymose system begins with a dichasial branch (F2, F2') and then continues with monochasial cymes (e.g. F3, F4)*

Cymose systems appear only from the first-order lateral branches. The cyme begins with a lateral shoot terminating in a flower (the terminal flower F1 in red in Figure 7.6(B) because it is initiated by the racemose meristem). Between this flower and its bracteole(s) (b in Figure 7.6(B)), a sympodial meristem (SM) forms, producing a lateral flower (F2 and F2') pushing the bracteole outward (red arrows on Figure 7.6(B)). The terminal flower bracteole (b) becomes the bract of the lateral flower (B). If there is only one lateral flower, the cyme is a *monochasium*, and if there are two lateral flowers, the cyme is a *dichasium*. Secondary sympodial meristems may form between flowers F2 and F2' and their bracteoles, and so on.

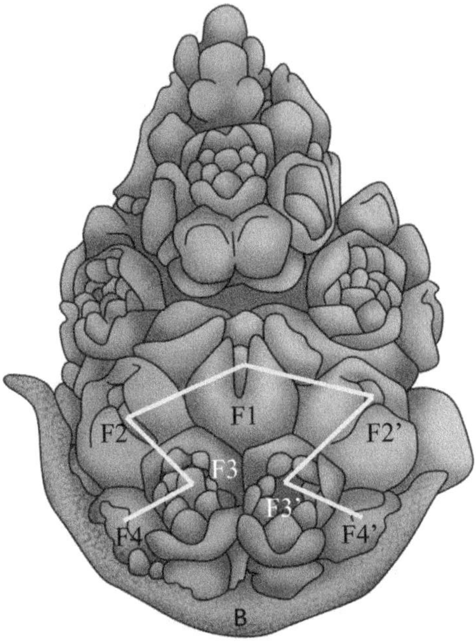

Figure 7.7. *SEM of terminal thyrse of Salvia verticillata (redrawn from Claßen-Bockhoff and Bull-Hereñu (2013)). F1-F2 and F2' = dichasial cyme and F2-F4 and F2'-F'4 = monochasial cymes. The leaves are opposite-decussate*

Different patterns of cymose branching are the consequence of the degree of branching and the number and position of consecutive meristems. Cymose inflorescences can be composed of either only a monochasial cyme (Myosotis, Boraginaceae), or an initial dichasial cyme followed by dichasia (Caryophyllaceae), or monochasia (scorpioid cyme (e.g. *Salvia verticillata*), helicoid cyme, drepanium, rhipidium).

7.2.1. *Racemes*

7.2.1.1. *Antirrhinum*

The inflorescence of snapdragon, *Antirrhinum majus* (Plantaginaceae), is a simple raceme with an apical meristem (IM) that does not form a terminal flower and produces floral meristems (FM) following a (3, 5) Fibonacci pattern similar to that of *Arabidopsis* (Chapter 4), with parastichies and orthostichies (Figure 7.8).

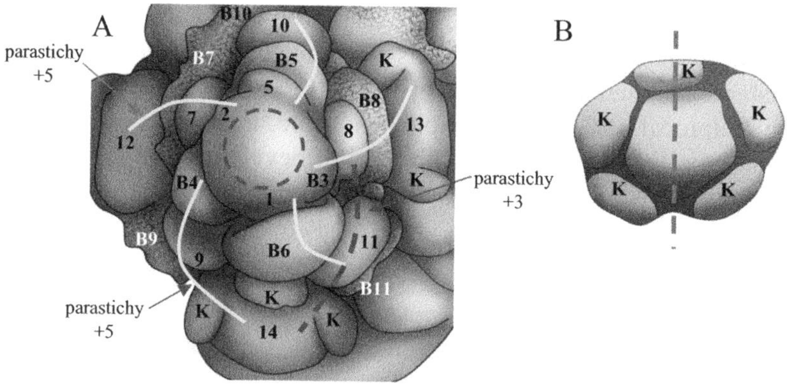

Figure 7.8. *(A) SEM of inflorescence apex of Antirrhinum majus: numbers correspond to the nodes, 1 being the youngest; Bn correspond to bract numbers; bracts B7-B10 have been removed to make floral meristems visible; K = sepals. In yellow, the five left parastichies; in red, one of the three right parastichies. Nodes 9 (14 – 5 = 9) and 11 (14 – 3 = 11) are located on the two parastichies passing through node 14; node 6 is located on the same pseudo-orthostichy as node 14 (14 – 8 = 6). (B) A. majus floral meristem at stage 4. Dashed line = axis of symmetry. Redrawn from Carpenter et al. (1995)*

Carpenter et al. (1995) defined the stages of development of the inflorescence of *A. majus*. At stage 0, tongue-shaped bract primordia are visible (nodes 1–5). At stage 1, floral meristems start being visible between their subtending bract and the central region (nodes 5–9). Floral meristems grow during stage 2 (nodes 9–11) and take on a pentagonal shape at stage 3 (nodes 11–13). At stage 4, first sepal primordia are clearly visible (nodes 13–15).

We hypothesized that four sources of inhibition influence the initiation of sepals in *A. majus*: the bract subtending the flower (θ_B = 270° = absolute angle of bract B), the two bracts located above the flower on the two parastichies passing through this flower (θ_{B-3} = 45°; θ_{B-5} = 135°) and the large adaxial bract (B-8) located on the pseudo-orthostichy passing through the flower. The bracts of the flowers above in the inflorescence are represented by inverted crescents. The inclination of these three upper bracts is consistent with the SEM photographs of Carpenter et al. (1995; Figures 7.8(A) and 7.9). The effective distances of the different bracts were chosen so as to reproduce the tepal arrangement observed on the SEM. The effective distances of the two lateral bracts need not be equal to produce the bilateral symmetry of the sepals ($R_{B\text{-}3} = 4$; $R_{B\text{-}5} = 3$; $R_{B\text{-}8} = 2$). Indeed, the reciprocal of Curie's principle is not true: the effects can be more symmetric than the causes. At the end of development, the flower of *A. majus* is strongly *zygomorphic* (*bilaterally symmetrical*; Figure 7.8(B)).

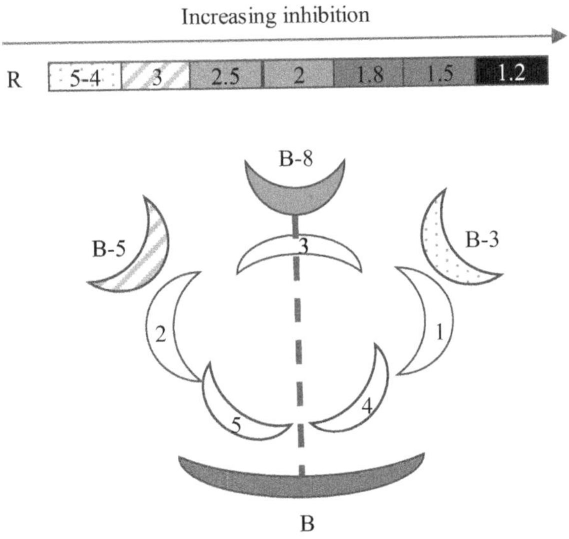

Figure 7.9. *Model of the floral meristem of Antirrhinum majus. Dashed line = axis of symmetry*

The influence of these bracts is due to contact pressure. Bull-Hereñu et al. (2022) noted that when the bracts in a racemose inflorescence are of a rigid configuration, they compress the young developing flowers. When the bract exerts extensive pressure, the floral primordium becomes squeezed and even flattened (FM14 on Figure 7.8(A)).

7.2.1.2. *Geranium robertianum*

Geranium robertianum (Geraniaceae) is a common plant throughout many of the temperate zones of the northern hemisphere.

Figure 7.10. *Geranium robertianum (Y. Alidina and S. Blaise)*

Its inflorescence is a botryoid (a raceme with a terminal flower) composed of racemes and cymes (= thyrse). The terminal flower (I) is axillated by four leaves (a, b, c, d): according to Eichler (1878), these four leaves are on the vegetative spiral, which then continues into the calyx of the terminal flower. Two lateral racemes, with terminal flowers (III and IV), are formed in the axils of leaves a and b (Figures 7.11(A) and (B)); a cyme (II) is formed in the axil of leaf c, while leaf d is *sterile* (no flower is formed in its axil).

For the modeling, leaves a, b, c, d were initialized following a Fibonacci spiral (divergence angle ~137°), with radial distances decreasing from a to d (Table 7.1): this determines the quinconcial initiation of the terminal flower I (Figure 7.11(C)).

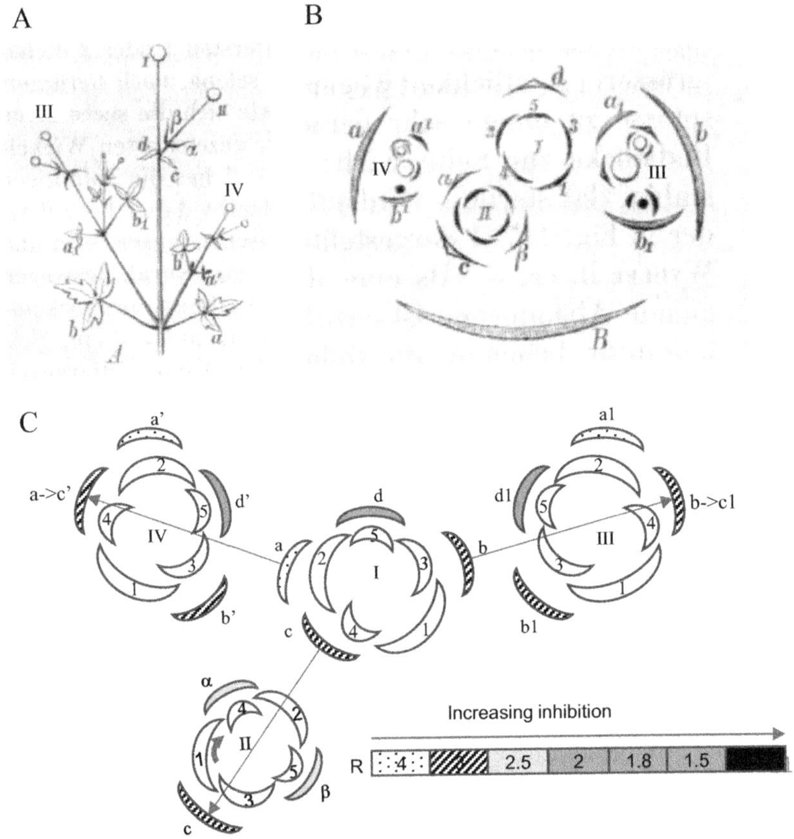

Figure 7.11. *(A) Morphological and (B) ontogenetic floral diagrams of Geranium robertianum (Eichler 1878). (C) Model of the partial inflorescence of G. robertianum*

Primordium number	Absolute angle (θ°)	Radial distance (R)
a	150	4
b	10	3
c	230	3
d	90	2

Table 7.1. *Plastochron ratios used to model floral organ initiation in Geranium robertianum*

A sympodial meristem (II) develops between leaf c (which becomes its bract) and flower I. Monopodial meristems develop between leaves a and b and flower I: they initiate the cauline leaves of racemes III and IV and the terminal flowers of these racemes, which have the same arrangement as that of flower I.

7.2.2. *Cymes*

In cymose systems, a sympodial meristem (e.g. SM4 on Figure 7.12) is initiated between the preceding flower (FM3) and one of its bracteoles (β3). It develops bracteoles and then floral organs: it will then be a floral meristem (SM4-> FM4).

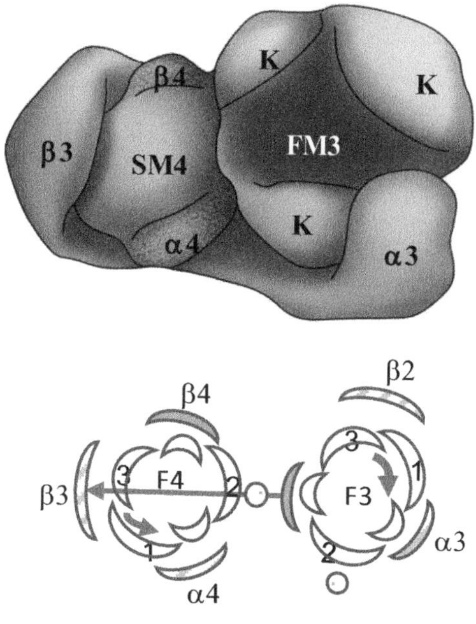

Figure 7.12. *Floral initiation in Petunia (redrawn from Souer et al. (1998)). On the left, a sympodial meristem (SM4) has initiated two bracteoles ($\alpha 4$ and $\beta 4$); on the right, the floral meristem (FM3, initiated before SM4) has already initiated two bracteoles ($\alpha 3$ and $\beta 3$) and three sepals (K; see Figure 7.26)*

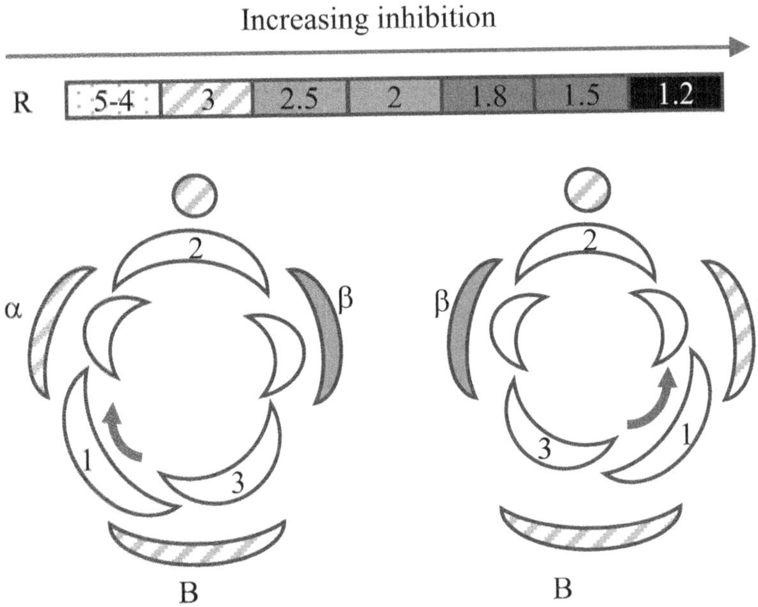

Figure 7.13. *The two possible directions of rotation of a quincuncial flower: clockwise and anti-clockwise*

As the floral diagram is classically represented with the bract "at the bottom" and the axis of the flower "at the top", bracteole β can be located on the right or left of the flower. The sepal K1 is diagonal-opposite β since the inhibition of the latter is of a higher level than α. *The position to the right or left of bracteole β determines the "direction of rotation" of the quincuncial flower: clockwise or anti-clockwise* (Figure 7.13).

In Figure 7.14(C), after the development of the terminal flower (F1: Figure 7.14(B)), two sympodial meristems develop between F1 and its bracteoles, which are pushed back along the axes shown by the arrows. The two primary bracteoles α1 and β1 become the bracts of the two lateral flowers F2 and F2'. The flower on the left is said to "*develop in α*", the flower on the right, in β. The lateral sympodial meristems initiate two bracteoles β2 and β2' whose position, to the right or left of the flower axis, determines the direction of rotation of F2 and F2'.

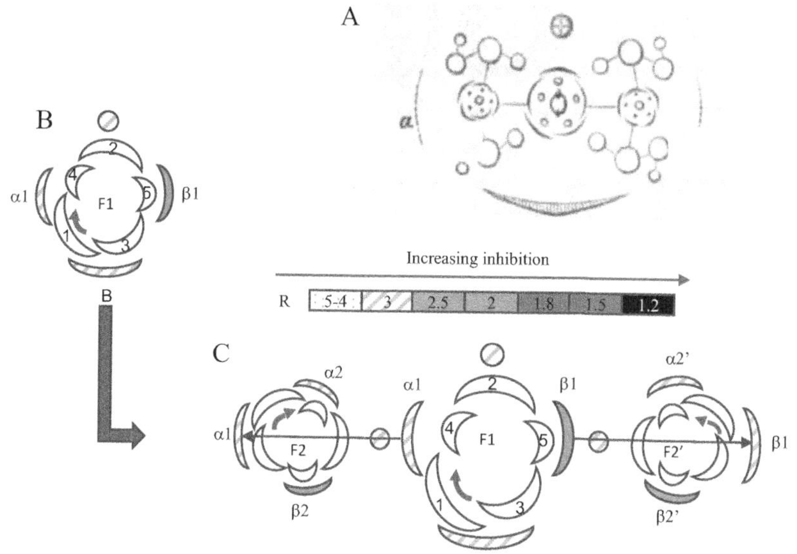

Figure 7.14. *(A) Diagram of the biparous cyme of Chenopodium album (Eichler 1878). (B) Modeling of the terminal flower. (C) First-order development of lateral flowers in a dichasium. Inhibition can change when bracteole becomes bract (e.g. β1)*

The construction method is as follows (Figure 7.15). Flower F2 should be viewed from behind the new location of α_1, which became its bract, toward its axis (dotted circle). Bracteole β_2 of F2 is located on the right of this axis. As a result, the calyx of F2 is *homodromous* to that of F1, that is, the "direction of rotation" of the two calyces is the same (clockwise). Flower F3 should be viewed from behind β_2, which became its bract, toward its axis. Bracteole β_3 is located to the left of that axis. As a result, its calyx is *antidromous* to that of F2, that is, the "direction of rotation" of the two calyces is opposite (clockwise and anti-clockwise).

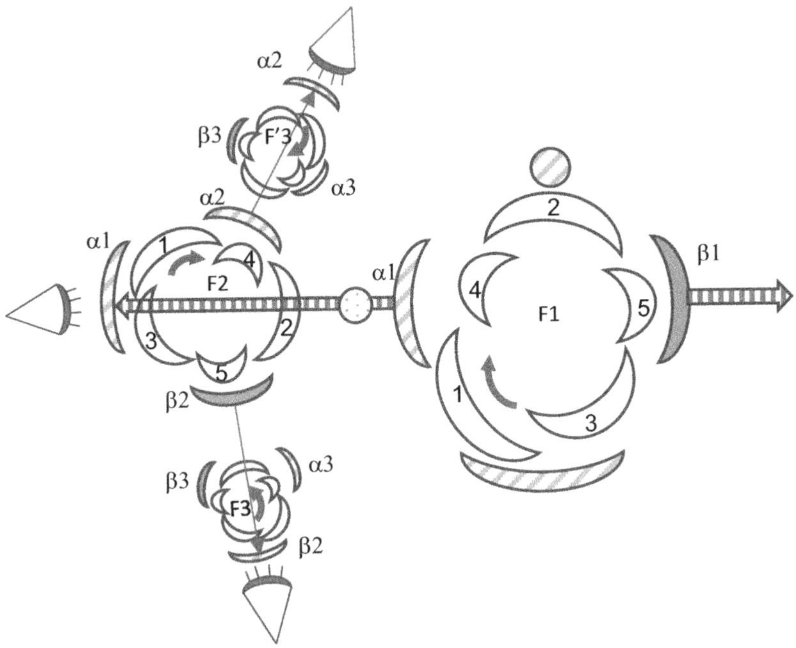

Figure 7.15. *Partial model showing construction method*

7.2.2.1. *Dichasium (biparous cyme)*

In Ranunculaceae, *Ranunculus acris* (Figure 7.16) and *Caltha palustris* (Figure 7.17), the secondary dichasium formed in α is more developed than that formed in β: their dichasium is said to be *favored in α.* In all other families (e.g. *Mirabilis jalapa,* Nyctaginaceae, Figure 7.19), it is the dichasium formed in β that is the most developed. In both cases, the favored bracteole axils a flower that is antidromous to the terminal flower (their direction of rotation is opposite), and the other bracteole axils a flower homodromous to the terminal flower.

7.2.2.1.1. Ranunculus acris

The inflorescence of *Ranunculus acris* (Ranunculaceae) initially forms a dichasium with a terminal flower F1 and two bracteoles that function as bracts (α and β) subtenting either a biparous cyme, on the left in Figure 7.16, when both F2 bracteoles are fertile, or a uniparous cyme, on the right, when one bracteole of F2'is sterile.

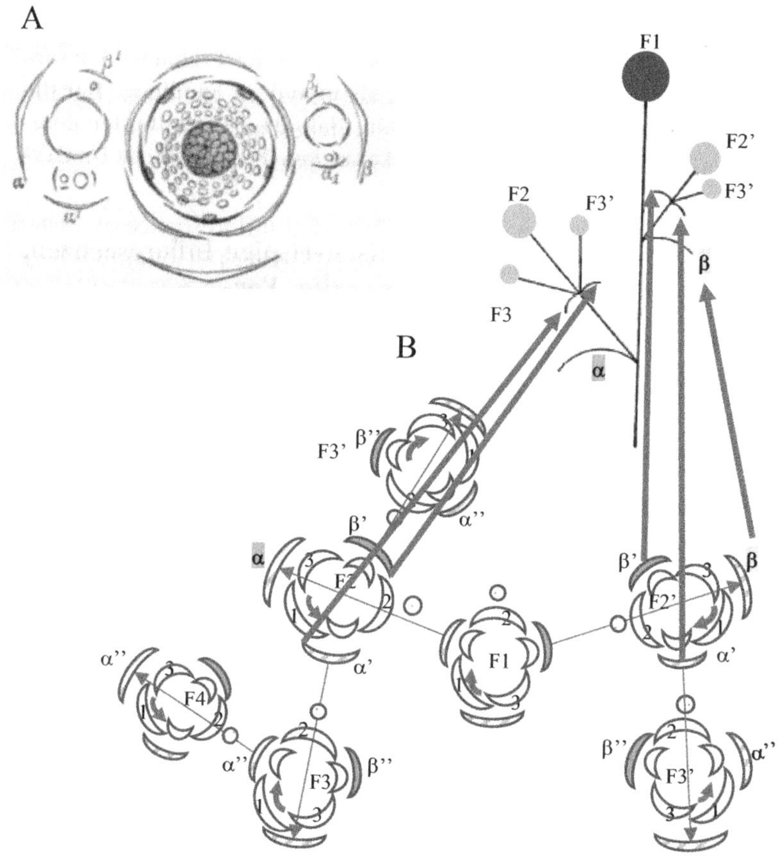

Figure 7.16. *(A) Ontogenetic groundplan of Ranunculus acris (Eichler 1878). (B) Morphological elevation and model of the inflorescence of R. acris*

There are several orders of flowers: the terminal flower (F1), which forms first and terminates the main axis; the primary lateral flowers (F2 and F2') of the dichasium, each subtended by one of the bracteoles of the main axis (α and β); the secondary flowers of the axillary cymes (F3 and F3'). A tertiary flower (F4) is formed in F3 bracteole α" axil (Figure 7.16(B)).

Sepals of flowers have a main quincuncial arrangement, which is determined by the two bracteoles α and β, in accordance with the model described in Figure 7.4(A).

7.2.2.1.2. Caltha palustris

Caltha palustris (Ranunculaceae), known as marsh-marigold or kingcup, is native to marshes, fens, ditches and wet woodlands in temperate regions of the Northern Hemisphere.

Eichler (1878) drew a diagram of a partial inflorescence of a lateral branch of *Caltha palustris* with a terminal flower (F1) and two lateral flowers (F2, F2'; Figure 7.17). The terminal flower has two bracteoles, α and β, in the axils of which the lateral flowers F2 and F2' arise. Flower F2 (in the axil of α) has one bracteole, while F2' (in the axil of β) has none. The K4 sepal of the F2' flower is adaxial.

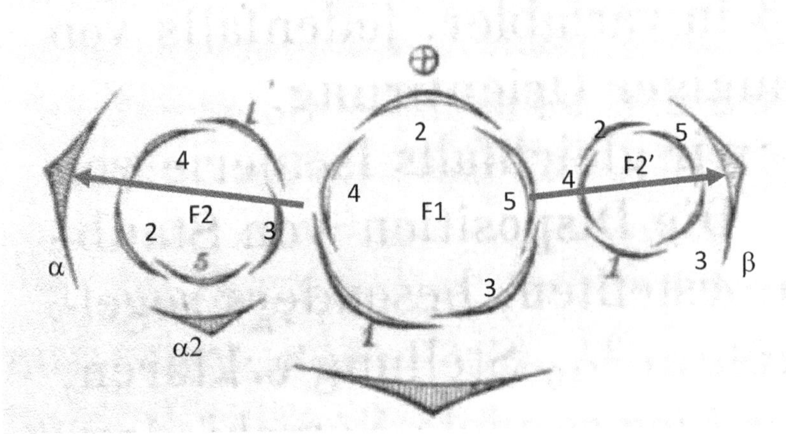

Figure 7.17. *Diagram of a partial inflorescence of C. palustris (Eichler 1878)*

Sepals of the terminal flower have a main quincuncial arrangement, which is determined by the two bracteoles α and β, in accordance with the model described in Figure 7.4(A).

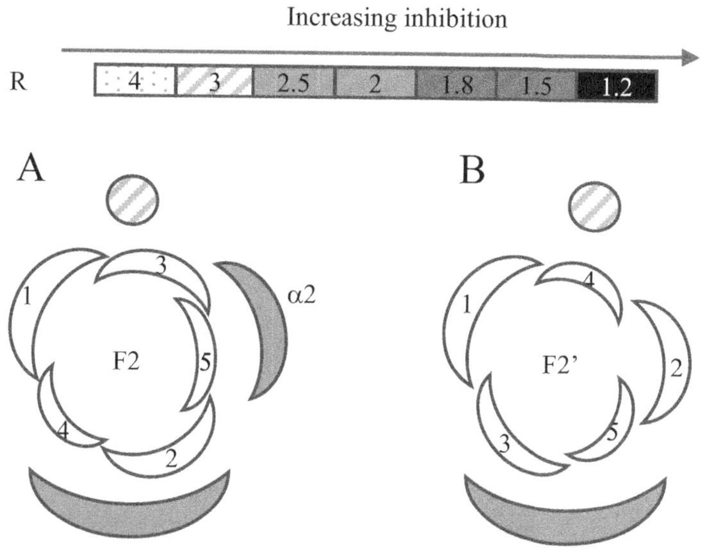

Figure 7.18. *Sepal initiation in Caltha palustris (A) in the F2 flower and (B) the F2' flower whose arrangement is of the Primulaceae type (K4 is adaxial). Circle = adaxial repulsion source*

Sepals in the lateral flowers of *C. palustris* cannot be in a main quincuncial arrangement (K2 adaxial) because they do not have two bracteoles. The first sepal of F2 and F2' is located on the dorsal side (and not on the ventral side as in the main quinconcial arrangement) away from the bract and the single bracteole for F2. Modeling of these arrangements requires an adaxial source of "repulsion" for F2 and one that is slightly offset for F2'. These sources can be located near the point of contact between the lateral flower and the terminal flower (dashed circle in Figure 7.18).

The terminal flower develops before the lateral flowers. Thus, the α and β bracteoles must already have played their role as a source of inhibition during sepal initiation in F1 before being pushed back by the sympodial meristems and becoming the bracts of the lateral flowers. We can believe that the adaxial and abaxial repulsions influencing sepal initiation in F2 and F2' are not caused by auxin concentration, but rather by contact pressure exerted on the one hand by the flower F1 and on the other hand by the bract α or β (see Chapter 5).

7.2.2.1.3. Mirabilis jalapa

Mirabilis jalapa (Nyctaginaceae), the "marvel of Peru", a commonly grown ornamental species of *Mirabilis*, was cultivated by the Aztecs for medicinal and ornamental purposes.

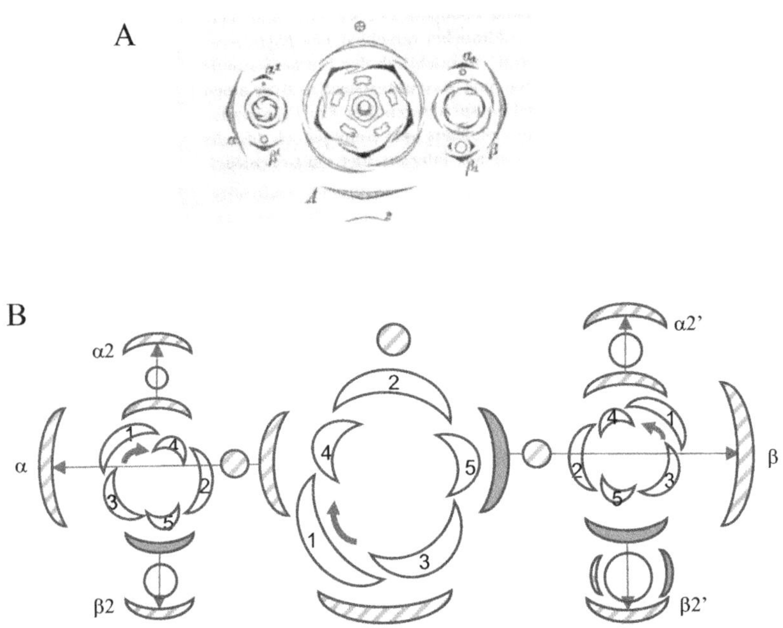

Figure 7.19. *(A) Inflorescence of Mirabilis jalapa (Nyctaginaceae) (Eichler 1878). (B) Model of a Mirabilis jalapa inflorescence*

The development in α is less important than in β, but nevertheless α remains fertile until the second order.

7.2.2.1.4. Chenopodium album

Chenopodium album is a fast-growing weedy annual plant extensively cultivated and consumed in India as a food crop known as *bathua*. The inflorescence of *Chenopodium album* (Amaranthaceae) is a dense ball (*glomerulus*) bearing numerous small white flowers. The first two lateral flowers are opposite, the bracteoles of the terminal flower forming a flat angle. On these branches an iterative cymose system with salient angles (i.e. less than 180°; Figure 7.20) is formed.

Figure 7.20. *(A) Inflorescence of Chenopodium album (Eichler 1878). (B) Model of a Chenopodium album glomerulus*

The inflorescence is dichasial up to F3 because all the bracteoles are fertile, after which only the β bracteoles are fertile and the inflorescence becomes monochasial.

7.2.2.2. *Monochasium (uniparous cyme)*

7.2.2.2.1. Scorpioid cymes or *cincinnus*

The *cincinnus* or scorpoid cyme (de Candolle 1827; in German Wickel) owes its Latin name (meaning “a lock of hair”) to its zigzag shape, the new lateral shoots appearing alternately on either side of the pseudo-axis of the inflorescence, the youngest flower being at the apex of the inflorescence (Figure 7.21(A)). This pseudo-axis is formed by the peduncles placed end to end (Figure 7.21(E)). Viewed from the side, it is rolled up, like the tail of a scorpion (Figures 7.21(B) and (D)). Scorpioid cymes are very frequent in dicots and rare in monocots.

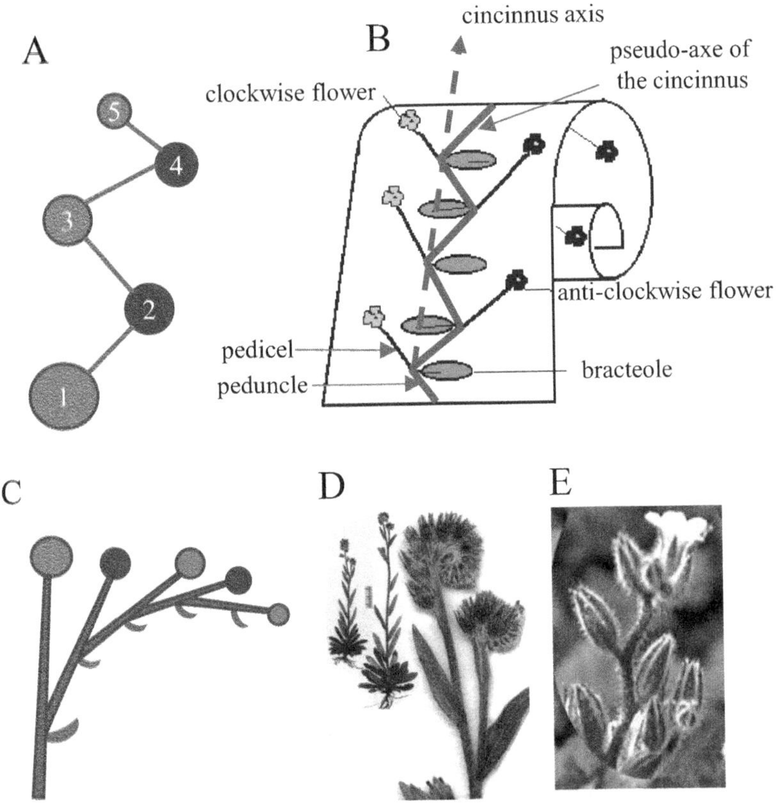

Figure 7.21. *Scorpioid cyme (A) Groundplan. (B) Three-dimensional representation (Blaise 2011). (C) Side view. Brown flowers come out towards the reader; blue flowers sink to the other side of the page. (D) Myosotis discolor with rolled up cyme (S. Blaise and Tison). (E) Front view of M. incrassata inflorescence (S. Blaise) showing a zig-zag pattern. The lateral flowers are already withered while the terminal flower is still open*

Once stretched (Figure 7.21(E)), the scorpioid cyme outlines the zigzag often mentioned in the literature. Even if the axis is almost rectilinear, the small axes being placed end to end form a pseudo-axis bearing flowers laterally on both sides. A scorpioid cyme cannot be confused with a raceme because in a scorpioid cyme the flower is on one side of the axis and the bract is on the other side, whereas in a raceme the bract is on the same side of the axis as the flower (Figures 7.21(B) and 7.27(C)). Bracts and flowers are located on two antidromous spirals.

7.2.2.2.2. *Salvia*

Salvia verticillata (Lamiaceae), the lilac sage or whorled clary, is a herbaceous perennial. The inflorescence of *Salvia verticillata* is a thyrse with opposite-decussate bracts, each axillating a dichasium of scorpioid cymes (Figures 7.22 and 7.23). The calyx is valved (gamosepaly), and the sepals K1, K5, K'1 and K3, K'3 fuse during development (Figures 7.24 (B) and (C)).

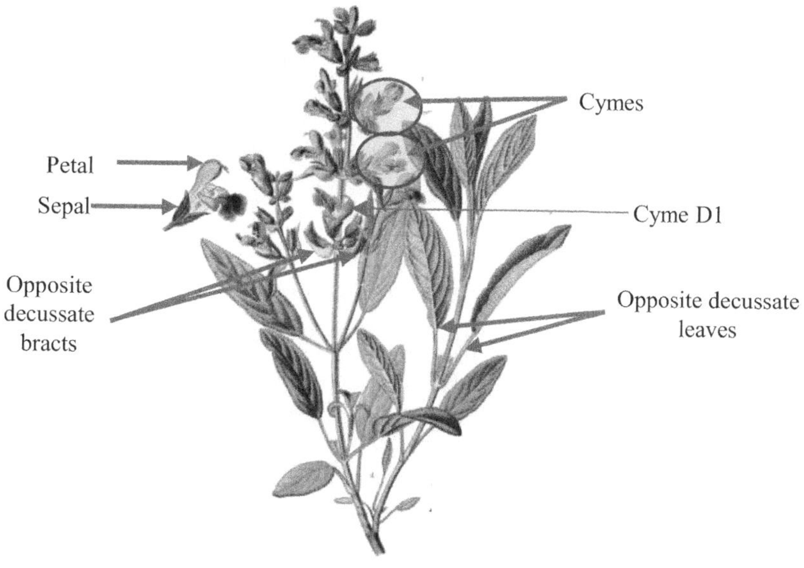

Figure 7.22. *Painting of Salvia officinalis by Köhler (1887)*

The arrangement of the sepals of *Salvia verticillata* is not quincuncial but valved (K5 adaxial) because the two bracteoles generate equal levels of inhibition (R = 3: note that Eichler did not call these bracteoles α and β). The fusion of organs is common in advanced groups of eudicots, particularly within calices and corollas, which are then "valved". The flower of *Salvia verticillata* is perfectly zygomorphic (bilateral symmetry; Figure 7.24(D)).

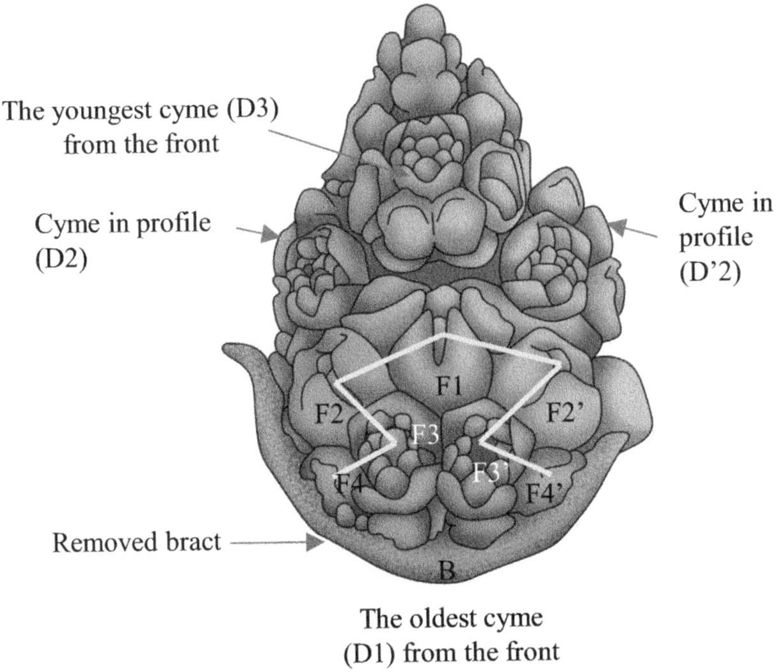

Figure 7.23. *Terminal thyrse of Salvia verticillata (redrawn from Claßen-Bockhoff and Bull-Hereñu (2013)) with opposite-decussate cymes. B = bract*

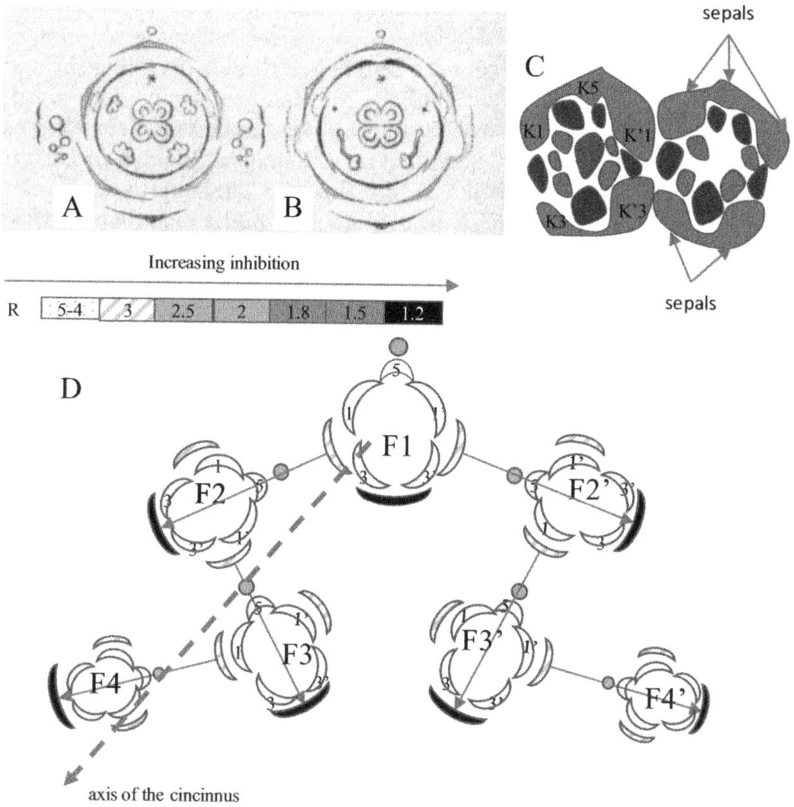

Figure 7.24. *Floral diagrams of Lamiaceae (Eicher 1875): (A) Lamium album and (B) Salvia officinalis with sepal fusion. (C) Drawing of Salvia verticillata showing flowers F3 and F3' from Figure 7.23 (Claßen-Bockhoff and Bull-Hereñu). Green = progressively fused sepals; red = petals; blue = stamens; purple = carpel. (D) Modeling of the biparous cyme of Salvia verticillata shown in Figure 7.23*

7.2.2.2.3. Atropa

Atropa belladonna (Solanaceae), commonly known as belladonna or deadly nightshade, is a toxic perennial herbaceous plant native to Europe, North Africa and Western Asia. As in most Solanaceae, the leaves are spiral and the main inflorescence is a thyrse terminated by a flower. The calyx of *Atropa belladona* has a quincuncial arrangement that becomes secondarily valved by partial fusion as in *Salvia* and *Lamium*.

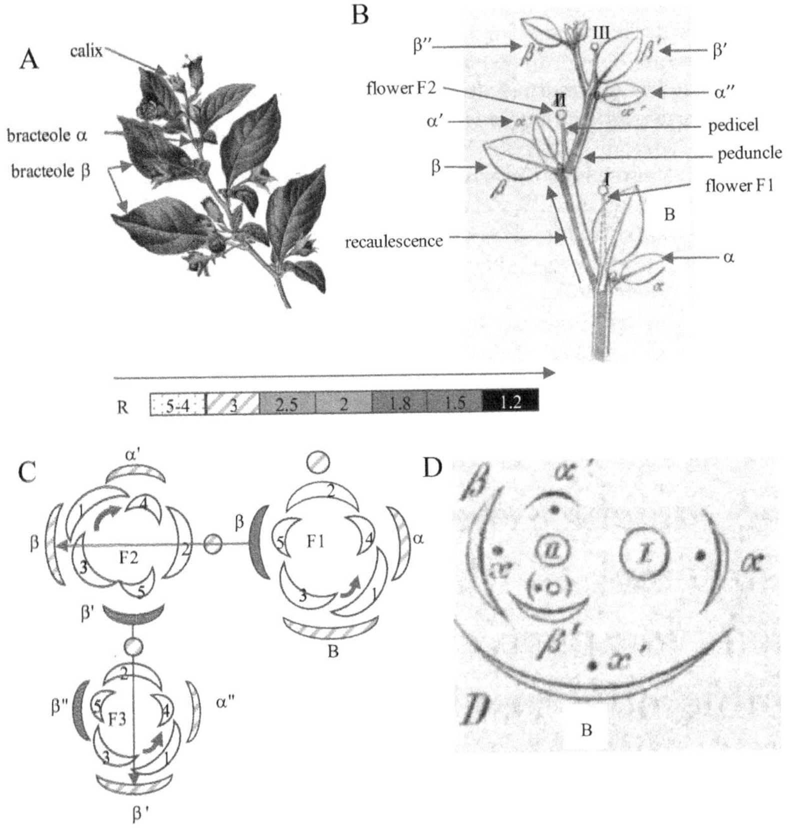

Figure 7.25. *(A) Painting of Atropa belladonna by Köhler (1897). (B) Elongation with recaulescence. Notice that the small axes placed end to end form a pseudo-axis (Eicher 1875). (C) Model of the partial scorpioid inflorescence. (D) Inflorescence diagram (Eichler 1875). x and x' are lateral buds axillated, respectively, by β and B*

The inflorescence of *Atropa belladonna* is a monochasial *cincinnus*. The diagrams by Eichler (1875; Figure 7.25(D)) show the inflorescence once the F1 and F2 flowers have developed, as well as the position of the F3 and F4 flowers. The elongation (Figure 7.25(B)) shows the inflorescence in a vertical direction. The displacement of bracteole β is such that at the first floral node, we find the bract of the inflorescence (B), as well as bracteole α of F1. Each successive node has bracteole α of the flower developing at that node, as well as bracteole β of the previous flower.

This migration of a leafy organ along an axis is called *recaulescence*. Here, the petiole of β fuses with the peduncle of the flower at the next node.

7.2.2.2.4. Petunia

Petunia (Solanaceae) is a genus of 20 species of flowering plants of South American origin. Well-known members of Solanaceae include tobacco, cape gooseberry, tomato, potato, deadly nightshade and chili pepper.

A day-by-day scanning electron micrograph series of a single developing *Petunia* (Solanaceae) inflorescence apex showed that a sympodial meristem emerges in the axil of a bract (e.g. SM3 in the axil of β2) and produces a new shoot consisting of two bracts (α3 and β3) and flower fm3 (Figure 7.26(B); Castel et al. 2010).

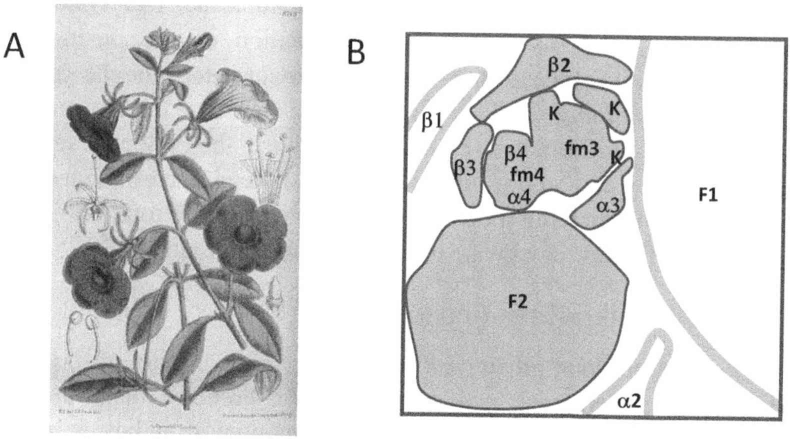

C

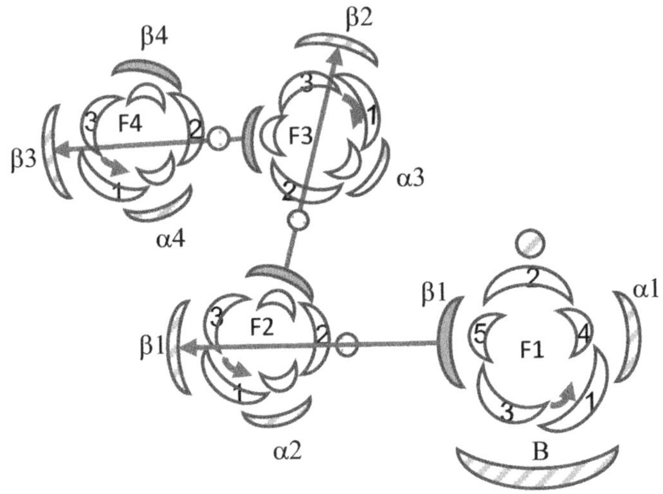

Figure 7.26. *(A) Painting of Petunia integrifolia by Kohler (1897). (B) Transverse section of the apex of a cyme of Petunia hybrida (redrawn from Castel et al. (2010)). (C) Modeling of the apex of a Petunia cyme*

The scorpioid cyme develops in β, which is usual for scorpioid inflorescences in most families except Ranunculaceae. This systematic development in β, as well as the alternation of the position of bracteole β with respect to the axis of the flower determine the general shape of a *cincinnus*, and in particular the fact that the flowers are successively antidromous (e.g. anti-clockwise then clockwise then anti-clockwise, and so on; Figure 7.26(C)). Indeed, the position to the right or left of bracteole β determines the "direction of rotation" of the quincuncial flower: clockwise or anti-clockwise (Figure 7.13).

7.2.2.2.5. Strelitziacaea (monocots)

Phenakospermum guyanense (Strelitziaceae) is a plant of the order Zingiberales (monocots) native to Suriname, French Guiana and the eastern Amazon River basin. Like most monocots, it has only one bracteole per flower. It is on the basis of its study that Kirchoff (2003) validates the application of Hofmeister's rule to the ontogeny of trimerous *cincinni*.

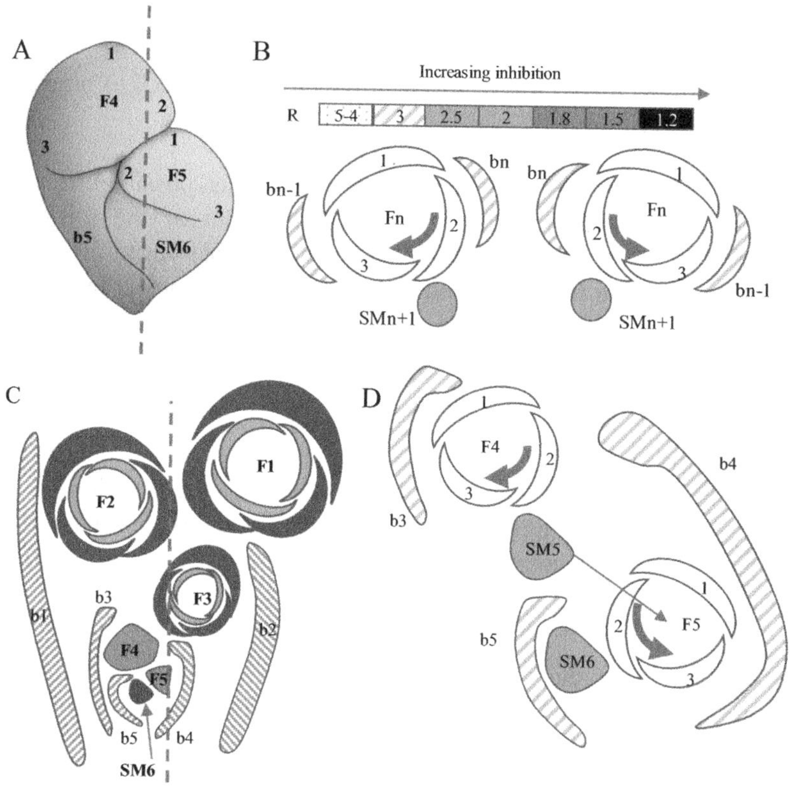

Figure 7.27. *(A) SEM of a partial cincinnus of Phenakospermum guyannense (redrawn from Kirchoff (2003)). The dashed line represents the central axis of the cincinnus. (B) Modeling of sepal arrangement in P. guyannense. The inhibition field is produced by two bracteoles (located at 150° and 330° in the flower rotating anti-clockwise), as well as by the sympodial meristem of the following flower (255° in the flower rotating anti-clockwise). (C) Partial diagram of a P. guyannese cincinnus at the initiation of F5. The dashed line represents the central axis of the cincinnus (redrawn from Kirchoff (2003)). (D) Modeling of the partial cincinnus of P. guyannese*

Flowers F1-F6 show the zigzag and are on one side of the axis when their bract is on the other side. Bracts and flowers are located on two antidromous spirals.

A first meristem initiated in the axil of the *cincinnus* bract produces a bracteole (b1) and the flower F1. Then, flowers are produced sequentially: a new sympodial meristem forms a bracteole (b2) and a flower (F2) in the axil of the bracteole b1. A new sympodial meristem (SM3 -> F3) then appears in the axil of bracteole b2 and so on (Figure 7.27).

In flower Fn, the first sepal forms as far as possible from its bracteole (b_n) and from the sympodial meristem of the following flower (SM_{n+1}; Figure 7.27(B)). The second sepal forms as far as possible from the first sepal and the bracteole of the preceding flower (b_{n-1}). The third sepal forms as far as possible from the first two. "*As far as possible*" *from two organs means close to their perpendicular bisector.*

Kirchoff claims that the position of the floral organs is determined by Hofmeister's rule. Our modeling shows that the position of the sepals depends on the positions of the two bracteoles (e.g. b4 and b5 for flower F5), as well as the following sympodial meristem (SM6).

7.2.2.2.6. Helicoid cymes

Viewed from the side, a helicoid cyme (or bostryx; in German Schraubel (meaning "loop")) can appear coiled like a scorpioid cyme. However, unlike in the latter where flowers are located alternately in front and behind a vertical plane (Figure 7.21(C)), flowers in a helicoid cyme are all located on the same curved line (Figure 7.28). Helicoid cymes are rare in dicots compared to scorpioid cymes and common in monocots.

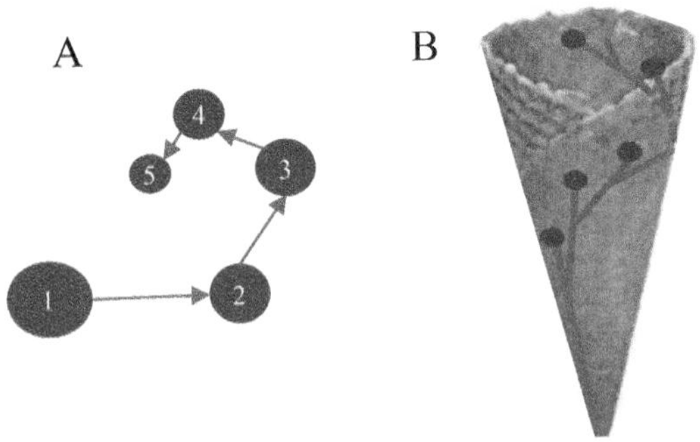

Figure 7.28. *Helicoid cyme. (A) Top view. (B) Three-dimensional representation (adapted from Blaise (2011))*

7.2.2.2.7. Beta lomatogona

Beta lomatogona is a species of wild beet in the family Amaranthaceae, native to Cyprus, Turkey, the Transcaucasus and Iran.

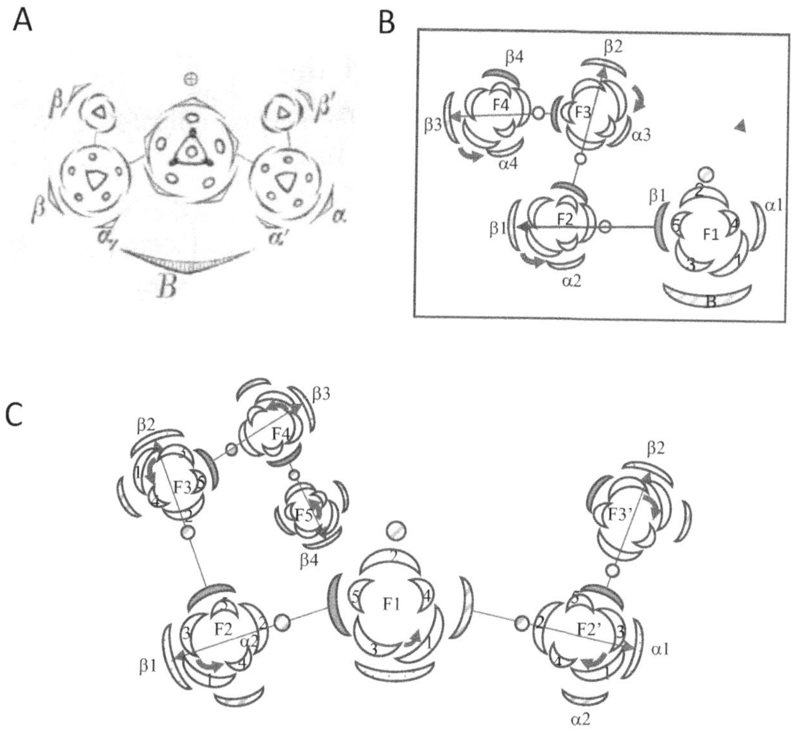

Figure 7.29. *(A) Diagram of the arrangement of the first flowers in a cyme of Beta lomatogona (syn. B. longespicata) (Amaranthaceae) (Eichler 1878). (B) Modeling of a scorpioid cyme. (C) Modeling of a biparous helicoid cyme in β (up to F5)*

In the scorpioid cyme in Figure 7.29(B), the β (fertile) bracteole of F3 lies to the right of the bract (looking from the bract toward the flower axis) unlike flower F2; as a result, flowers F2 and F3 are antidromous (Figure 7.26). In the helicoid cyme in Figure 7.29(C), the β (fertile) bracteole of F3 lies to the left of the bract (looking from the bract toward the flower axis) like flower F2; as a result, flowers F2 and F3 are homodromous. The side of the bracteole β alternates from flower to flower in scorpioid inflorescences and is the same in helical inflorescences. Therefore, two successive flowers will be antidromous in scorpioid cymes and homodromous in helicoid cymes.

7.2.2.2.8. Leucojum aestivum

In the tribe Galantheae (Amaryllidaceae, monocotyledons), flowers can be solitary (e.g. *Leucojum vernum* (the spring snowflake) and *Galanthus nivalis* (the snowdrop)) or form helical inflorescences (e.g. *Leucojum aestivum* (the summer snowflake)).

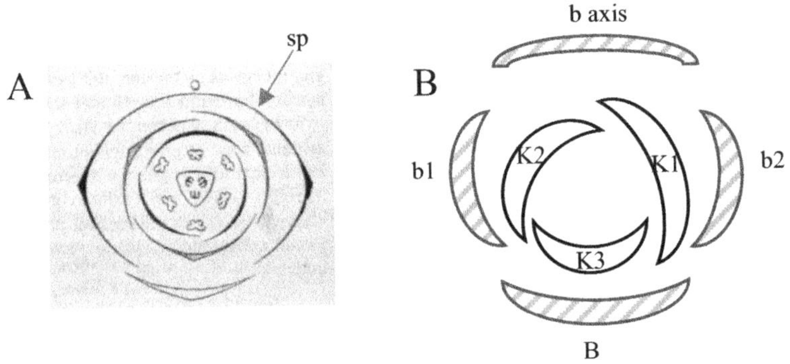

Figure 7.30. *(A) Floral diagram of Galanthus nivalis (from Eichler (1875)); sp = spathe. (B) Modeling of the G. nivalis flower ($\theta_B = 270°$, $\theta_{axis} = 90°$, $\theta_{b1} = 190°$, $\theta_{b2} = 350°$)*

In *Galanthus nivalis*, the floral bud is enveloped by a spathe (Figure 7.30(A)), which, according to Eichler (1875), is an amplification of a bikeeled adaxial bracteole (Figure 7.3(A)) and resembles two fused leaves. The spathe encloses the flower bud and forms a gutter. It is modeled here by two lateral zones of inhibition (b1 and b2) corresponding to the particular kinds of bracteoles and a dorsal zone (b_{axis}) corresponding to the fused part (Figure 7.30(B)). Thus, sepal K3 is abaxial and the flower is approximately ternary symmetric.

In *Leucojum aestivum*, the terminal flower of the double helicoid inflorescence forms in the same way as the single flower of *Galanthus nivalis* (Figure 7.30(A)). Lateral flowers, however, have only one bracteole, which is typical for monocots. The single bracteole is in a lateral position (configuration 3C, Figure 7.31(B); see Walch and Blaise (2022b)).

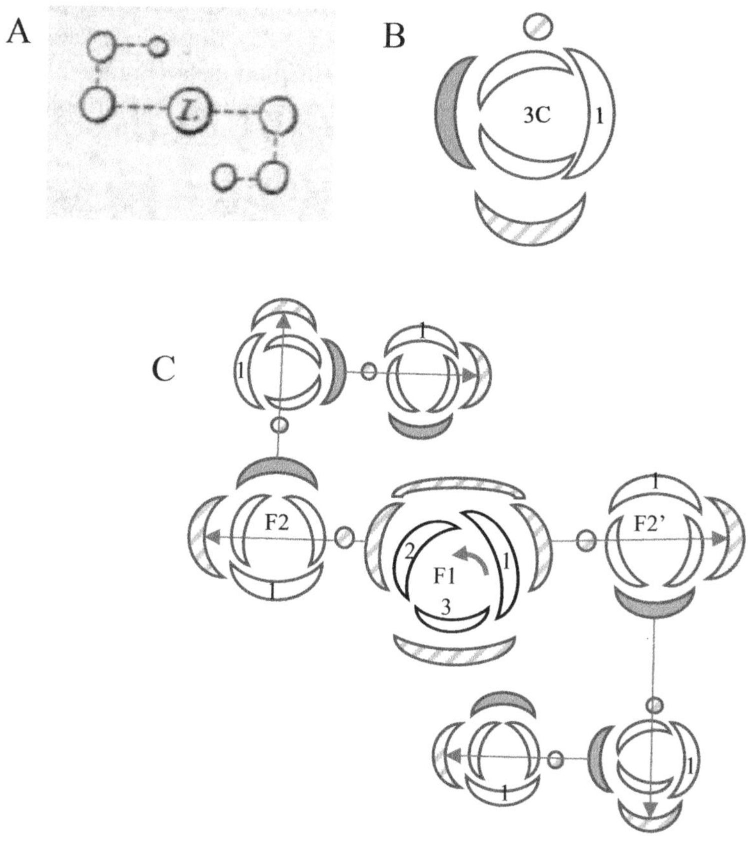

Figure 7.31. *(A) Diagram of the double helicoid inflorescence of Leucojum aestivum (Eichler 1875). (B) Modeling of the calyx of a lateral flower of L. aestivum. (C) Modeling of the double helicoid cyme of L. aestivum. F1 is the terminal flower*

The double helicoid inflorescence begins with a dichasium on either side of the terminal flower. Within each helicoid cyme, successive flowers are homodromous, which corresponds to an identical position of the bracteole with respect to the floral axis (Figure 7.31(C)).

7.2.2.2.9. Rhipidium

There are two other types of uniparous cymes: rhipidium (in German Fächel) or hand fans (Figure 7.32) and drepanium or sickles (Figure 7.33), the vernacular names corresponding to their shape in the vertical plane. According to Eichler (1875), rhipidia develop in the axil of the adaxial bikeeled bracteole in monocots (Figure 7.32(C)). This interpretation is disputed by those who believe that a rhipidium is not a cyme but a raceme (Hrabovsky 2019).

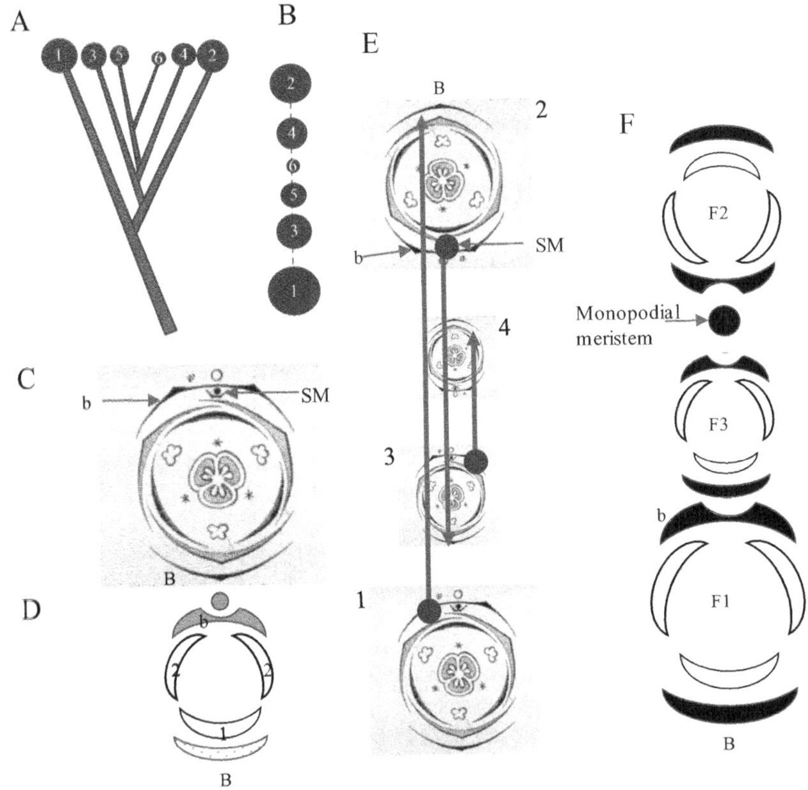

Figure 7.32. *The interpretation of a rhipidium as a cyme according to Eichler (1875) (A) in a vertical (B) and horizontal plane. (C) Floral diagram of Iris pseudacorus (Eichler 1875). (D) Modeling of the calyx. (E) Modeling of the rhipidium. (F) A partial distichous raceme with bikeeled adaxial bracteoles (redrawn from Remizowa et al. (2013)). SM = sympodial meristem. b = bracteole. B = bract*

For Eichler (1875), rhipidia are cymes, whereas for Remizowa et al. (2012, 2013), racemes in monocots can also form rhipidia. Indeed, the flowers of *Iris pseudacorus* (Iridaceae) are initiated from sympodial meristems (Figure 7.32(E), red circles), while those in Figure 7.32(F) are initiated from a monopodial meristem.

7.2.2.2.10. Drepanium

Juncus bufonius ("toad rush", Juncaceae) is a monocot with an involucral bikeeled adaxial bracteole ("g" in Figure 7.33(C)). Inside this involucrum are bracteole "a" and two involucral leaves "x" and "y". For Eichler (1875), this is an example of a drepanium (in German Sichel).

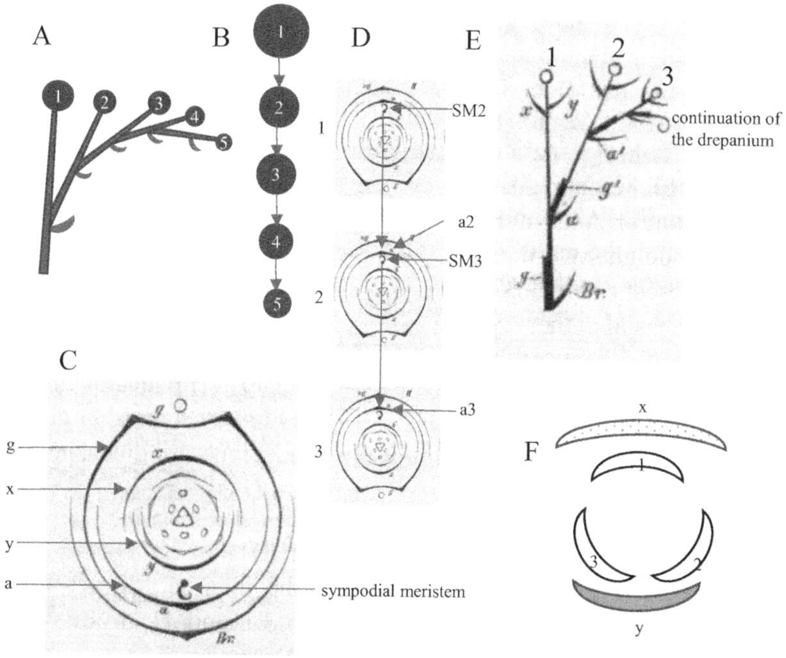

Figure 7.33. *A drepanium seen in a vertical (A) and horizontal plane (B). (C) Floral diagram of Juncus bufonius (Eichler 1875). (D) Modeling of the inflorescence. (E) Elongation of the inflorescence in J. bufonius (Eichler 1875): the bikeeled bracteole forms a gutter around each lateral branch. (F) Modeling of the flower. Br = bract; g = bikeeled bracteole; a = ventral bracteole; x, y = involucral leaves*

Viewed from the side (Figure 7.33(A)), the inflorescence resembles a scorpioid cyme, but seen from above (Figures 7.33(B) and (D)), the flowers are strictly aligned.

The arrangement of the calyx, with adaxial initiation of the first sepal, requires that the inhibition exerted by the ventral leaf (y) be greater than that exerted by the dorsal leaf (x), which corresponds to the observed distances of these two leaves on the floral diagram (Figures 7.33(C) and (F)). This pattern is consistent with the initiation of a trimerous perianth determined by two distichous involucral leaves (x and y).

7.2.3. *Involucra and spathes*

7.2.3.1. *Salicornia europaea*

Salicornia is a genus of succulent halophytic (salt tolerant) plants from the family Amaranthaceae. *Salicornia* species grow in salt marshes, beaches and mangroves and are native to North America, Europe, Central Asia and southern Africa. Common names for the genus include glasswort, pickleweed, picklegrass and marsh samphire. The inflorescence of *Salicornia* is a thyrse with opposite-decussate axillary dichasial cymes.

In *Salicornia*, the bracteoles fuse together to form sheaths around the main axis (Figure 7.34: I1, I1'; I2, I2'). Flowers appear in groups of three (e.g. F1, F2, F2'), forming opposite-decussate dichasia (D1, D2, D2', D3).

The initiation of a dichasial cyme (e.g. D1) in *Salicornia* is as follows. Flower F1 arises in the axil of the involucral bract I1. A sympodial meristem then appears between F1 and I1, from which flowers F2 and F2' develop (Figure 7.34(C)).

F2 and F2' axis is oblique with respect to the inflorescence axis as a result of the influence of contact pressure with flower F1, the involucral leaf (I1) and its lateral fusions (S1, S1').

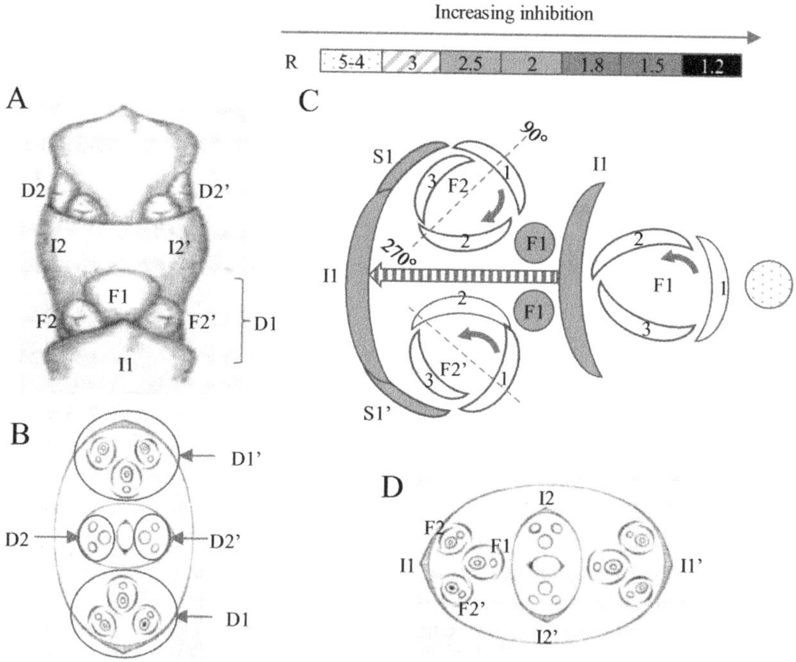

Figure 7.34. *Diagrams of the inflorescence of Salicornia europaea (Eichler 1878): (A) Front view. (B) and (D) top views. (C) Modeling of dichasium D1. Dashed line = F2 and F2' axis; I1, I1', I2, I2' = involucral scales; D1, D2, D2', D1' = dichasial cymes; F1, F2, F2' = flowers in the D1 cyme; 1, 2, 3 = sepals; S1, S1' = lateral fusions. D1', I1' are on the back side of the front view*

7.2.3.2. *Frankenia pulverulenta*

Frankenia (Frankeniaceae, Caryophyllales) is a genus of salt (halophytic) or drought (xerophytic) tolerant shrubs, subshrubs or herbaceous plants.

The inflorescence is a dichasium terminated by flower F1; lateral branches also form dichasia enclosed within tetramerous involucra. Only the a and b bracteoles enclosing flower F1 are fertile: they axillate flowers generated from sympodial meristems. Cymes preferentially develop in b (Eichler 1878; Figure 7.35(B)).

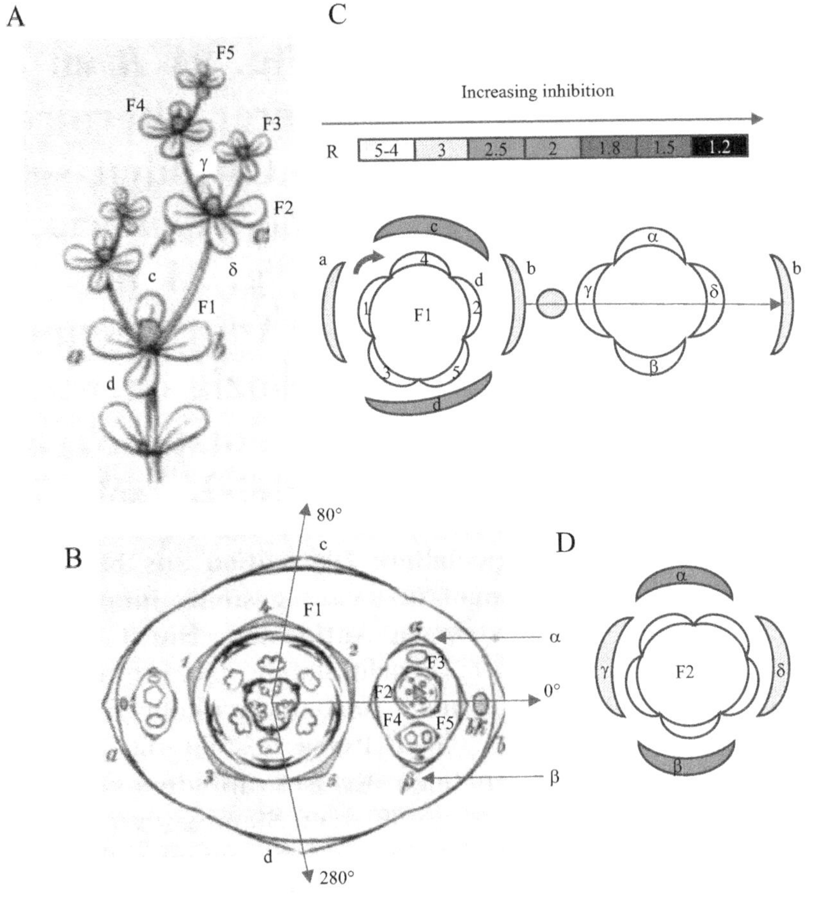

Figure 7.35. *(A) Elongation and (B) diagram of the inflorescence of Frankenia pulverulenta (from Eichler (1878)). Corresponding flowers are shown on each figure. (C) Initiation of the terminal flower F1 and the lateral involucrum. (D) Initiation of the lateral flower F2*

The main axis bears a few pairs of decussate leaves (the last one is visible at the base of Figure 7.35(A)), then a terminal flower that develops within an involucrum of four leaves (a, b, c, d) (Figure 7.35(B)). As Eichler (1878) depicted with accuracy, leaves c and d are not exactly opposite. Modeling shows that this arrangement is necessary for a Primulaceae-type pentamerous arrangement with an adaxial K4 of flower F1 ($\theta c = 80°$; $\theta d = 280°$; Figure 7.35(C)).

The initiation of the secondary involucrum (α, β, γ, δ) requires contact pressure exerted laterally by leaf b and flower F1.

7.2.3.3. *Nyctaginaceae*

The family Nyctaginaceae belongs to the order Caryophyllales. The best-known species in this family is bougainvillea (*Bougainvillea spectabilis*), which was brought to France from Brazil by Louis-Antoine de Bougainville in 1769. *B. spectabilis* has small white flowers that are surrounded by colorful bracts, often mistaken for petals.

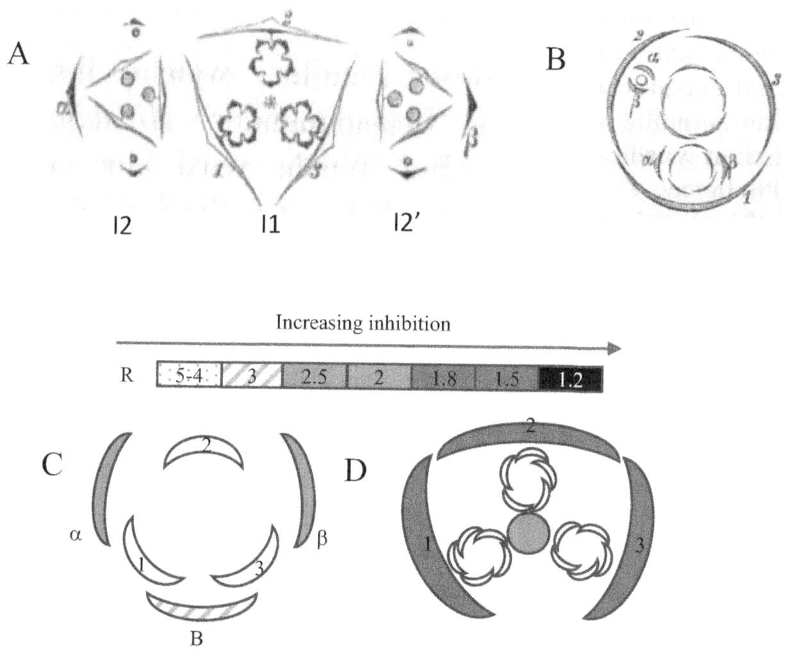

Figure 7.36. *(A) Inflorescence diagram of Bougainvillea spectabilis (Eichler 1878). (B) Inflorescence diagram of Anemone ranunculoides (Eichler 1878). (C) Initiation of a trimerous involucrum. Numbers are leaves. (D) Initiation of the pentamerous flowers in I1. The central aborted flower is represented by a circle*

The inflorescence first forms a trichasium. The central flower of I1 (Figures 7.36(A) and (D)) is aborted (star on Eichler's diagram,

Figure 7.36(A)). This trichasium is enclosed by two bracteoles, α and β, which allow the development of a trimerous involucrum (Figure 7.36(C)).

The lateral involucra (I2 and I2') develop in the axil of the two bracteoles α and β and are identical to the central involucrum. The perigon (undifferentiated tepals) is pentamerous of convolute aestivation and strongly gamophyllous (tepals are fused).

The ternary symmetry of each involucrum produces a three-flowered cyme. The axial arrangement of the two sources of repulsion (an involucral leaf and the central flower) leads to a pentamerous arrangement of the tepals; bilateral symmetry is at the origin of its convolute aestivation (Figure 7.36(D)). This repulsion is probably the consequence of contact pressure exerted by the involucral leaf and the central flower on the other flowers.

In *Anemone ranunculoides*, the central flower develops normally (Figure 7.36(B)).

Another Nyctaginaceae, *Mirabilis violacea* (syn. *Oxybaphus violaceus*), from South America, forms a pentamerous involucrum that can be mistaken for sepals as it encloses the flower in the manner of sepals (Ronse de Craene 2022).

The arrangement of this involucrum is quincuncial due to the influence of the α and β bracteoles (Figure 7.37(B)). Three flowers are initiated within this involucrum. The central flower (F1) is influenced by the five involucral leaves, which must be symmetrical angularly and radially, so as to produce a convolute aestivation. Two lateral flowers (F2 and F3) are initiated in the axils of the first two involucral leaves (1 and 2). The organ arrangement of these two flowers is determined only by the leaf in the axil of which they form and the central flower. This axial symmetry also gives them convolute aestivation.

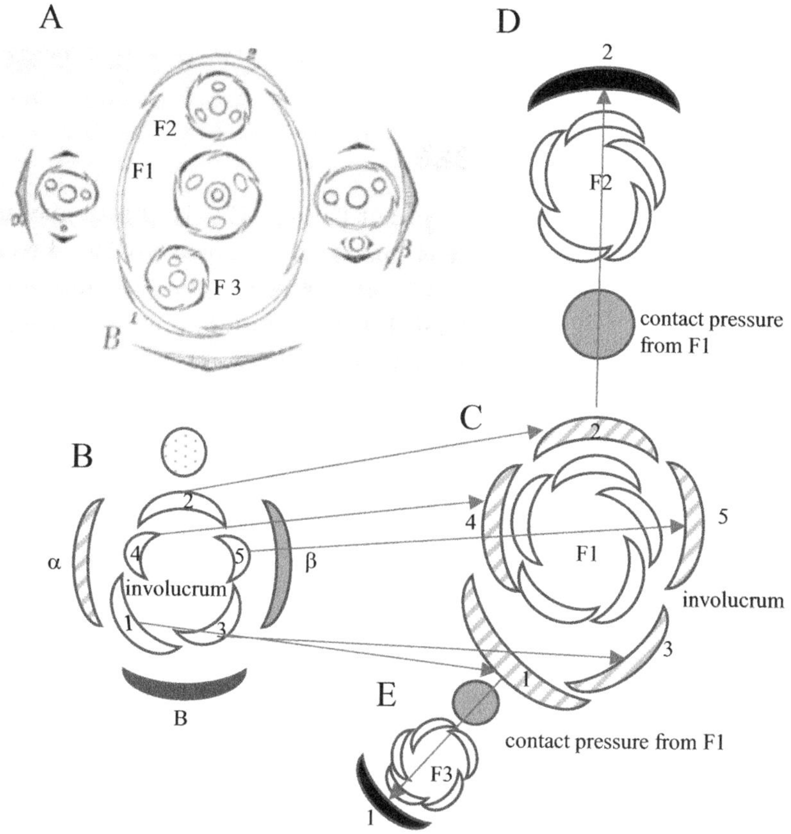

Figure 7.37. *(A) Inflorescence diagram of Mirabilis violacea (cymose lateral shoot; Eichler 1878). (B) Initiation of the pentamerous involucrum (numbers are leaves) and (C) the flower F1 inside the involucrum, (D) flower F2 and (E) flower F3*

The repulsive influence of the central flower and the involucral leaf could come from the contact pressure they exert on the lateral flowers.

The inflorescence develops in α and β, with two lateral quincuncial involucra within which three convolute flowers arise. Bracteole β, although younger than α, is larger because the cyme is more developed on the branch subtended by β.

Conclusion

Give me the plastochron ratios, I will draw you a flower.

C.1. From the pre-scientific stage to mathematization

For Gaston Bachelard (1938), physics developed over two great periods. The first period, in classical antiquity and the Renaissance (the 16th, 17th and even the early 18th century), was "pre-scientific". The second period, with the development of scientific methods, began in the late 18th century and flourished in the 19th and 20th centuries. In the pre-scientific period, botany adopted a rather colorful and somewhat anthropomorphic lexicon, which was actually often quite practical because it was perfectly suited for naturalist descriptions. These expressions pervade the botanical writings of Bravais and Bravais.

The bract is called the "*mother-leaf or mother-bract*", at the "*armpit*" of which a branch arises; it is unclear whether this branch is derived from the "*vital node*" that is protected by the bract or on the contrary if the bract plays a "*role in the generation of the axillary bud*". The latter seems more likely (Bravais and Bravais 1837a).

> When examining the shape of two bracts on a binodal peduncle, we very often see a great difference between

For a color version of all the figures in this book, see www.iste.co.uk/walch/phyllotaxis.zip.

> the upper and the lower bract depending on *whether they give rise to a new peduncle or not*. When these two nodes are more or less equally *fertile*, these organs generally differ very little; the upper bract is sometimes slightly less developed than the lower one (*Ranunculus*, *Aquilegia*, *Potentilla*); but if one of these two nodes is habitually *aborted*, the corresponding bract is often much less developed than the other one. This can be observed in the unilateral spike of *Silene quiquei ulnera* and other *Silene*, in *Helianthemum glutinosum*, in the sarmentose stem of *Geranium molle* and *pyrenaicum*, where one of the twin leaves is large and bears two stipules whereas the other is reduced to a small scarious bract [...].
>
> In *Saxifraga sannetosa*, the presence of the first node is only indicated by a very narrow reddish bracteole. In *Sedum stellarum* and *reflexuni*, [and] *Echeveria grandiflora* the *sterile* lower leaf rapidly decreases in size and is reduced to an abscised piliform organ, of which there is no trace on the upper part of the stem. In many Solanaceae, where the *mother-bract* fuses with its axillary peduncle up to the height of the lower floral bract of this peduncle, we can easily see that the latter is smaller and more often abscised (belladonna, *Physalis ventricosa*); it is completely *aborted* in *Datura stramonium*, which does not prevent its axillary bud from developing into a leafy branch [...].
>
> Leaves can also be fused with the peduncle that arises at their *armpit*: this fusion is extreme in cymes of *Paliurus aculeatus* and *Bignonia catalpa*, the leaves growing up to the next node (Bravais and Bravais 1830).

In Eichler (1875), the vital node that gives rise to a flower is called a blastema (the germ) from the name given to the "plant embryo" by Nysten (1824).

Eichler often used the term "sterile" but seldom the term "fertile". This is how he describes *Atropa belladonna* (Figure 7.25): "on the stem, which is actually sympodial, there are always two leaves side by side: one is larger, the bract of a fully developed sympodial flower, and the other one is smaller: its lower *sterile* bract". The dichasium of *Ranunculus acris* is described by Eichler (1878) as "Förderung aus α" meaning "promoted from α" or "*favored* in α", the development in α being greater than in β.

The "vital node" became the "meristem", a term coined by the Swiss botanist Carl Wilhelm von Nägeli (1858) from the Ancient Greek μερος (meros) meaning "that divided into part".

In modern terminology, comprehensively illustrated by Ronse de Craene (2022), bracts and bracteoles are no longer "mother-leaves" and become subtending. The English language distinguishes axilla from armpits. Bracts and bracteoles can still be abortive. The term fertile is limited to stamens and carpels, and it is no longer used in relation to the emergence of lateral branches. Flowers are enclosed by bracteoles. In Ronse de Craene's comprehensive overview of floral diversity, the two bracteoles are seldom described as being unequal in size and never differentiated in terms of Eichler's α and β.

As for phyllotaxis, the modern theory based on ontogeny was formulated by Douady and Couder (1996a, 1996b) who modeled the effect of an inhibition or repulsion on the arrangement of many meristems and inflorescences in Asteraceae (sunflower; detailed by Couder 1998).

Mathematical phyllotaxis provides a theoretical basis to the ideas of Eichler (1875) regarding the influence of bracteoles on sepal arrangement. This giant in the field of botany would have liked his "Blüthendiagramme" to be a general theory of flower structure, but he lacked the tools to develop such a theory. Indeed, he admitted to having no theoretical explanation for the effect of bracteoles on the position of sepals: "Finally, I find no possibility of explaining in a *purely mechanical* way these different cases, as Hofmeister tried to do. But delving into this subject would take us too far, as the reader must have had the impression [...]. Different cases are easy to describe, but as difficult to explain as why one plant has round leaves and another triangular leaves" (Eichler 1875, p. 32). Our model of phyllotaxis explains these different cases in a "purely mechanical" way, thus responding to the challenge raised by Wilhelm Eichler 150 years ago (Chapter 7).

Our model seems particularly simple considering the diversity of plants and the complexity of the underlying biochemical processes. One editor told us: "it is clear that sometimes rearrangements of phyllotaxis (and symmetry in case of transitions between spiral and whorled phyllotaxis) may occur without a change in organ identity and in the 'inhibition potential'. They result mainly from the developmental noise – through accumulation of small deviations from accurate positioning of primordia in a history of the apical meristem. The whole picture is therefore much more complex than that drawn on the basis of the plants selected at random in this work". A reviewer of one of our manuscripts wrote: "We doubt any model is able to reproduce the full diversity of forms found in Nature and this has little to do with the number of parameters used".

On the contrary, we believe that our model is capable of accounting for all the diversity in plants, as well as synthesizing a set of highly complex microscopic phenomena. The "inhibition potential", irrespective of its biochemical or mechanical basis, drives the "whole picture". In physics, the laws of thermodynamics describe macroscopic phenomena, such as the evolution of entropy, based on a microscopic level, in this case the arrangement of an extremely large number of molecules. The macroscopic effects of the phyllotaxis engine – the ability to diversify but also the stabilization of well-identified species and the optimization of compaction within meristems – have led to its evolutionary success.

C.2. Spirals and whorls

The type of "main" quincuncial calyx (with K2 in an adaxial position) is initiated along a pseudo-whorl, that is, a spiral that mimics a whorl. This is the reason why we can number sepals sequentially, the first two being external, the third being under the first and above the fifth, and the last two being internal. In a whorled calyx, it is impossible to number sepals since they are all identical. However, in a quincuncial arrangement, the organs, whose divergence angle is ~137° at initiation, are equidistant on the circle, that is, with a divergence angle of ~144°, at anthesis.

Generations of botanists have until now been unable to explain this complexity.

The founding father of phyllotaxis, Charles Bonnet (1754), with the help of mathematician Jean-Louis Calandrini, classified the main leaf arrangements in plants: whorled, spiral, but also quincuncial, the most frequent arrangement in the 125 species he examined. In this case, as an approximation, the sixth leaf is above the first one, but its distance from the first leaf is such that "it would not impede its function". Then, examining an apricot tree, Bonnet noticed that the sixth leaf was not exactly above the first, nor was the seventh above the second, and so on, but that leaves form a spiral that turns around the stem. This is "pleasing" as it minimizes leaf overlap, which can be considered the "ultimate goal" of a quincuncial arrangement. Better still, this particularity or "declination" can be observed in other species.

Schimper (1830) and Braun (1835) discovered that the Fibonacci sequence is at the heart of spiral arrangements: in a quincuncial arrangement, five leaves are required to complete two turns, thus the angle between two successive organs has to be 2/5 of a turn, that is, 144°. If we continue along the Fibonacci sequence, eight organs would be required to complete three turns (3/8 = 135°) and 13 organs would be required to complete five turns (5/13 ~ 138°), etc. With this approach, the divergence angles take on a *rational value*, making these arrangements spiral with perfectly vertical orthostichies.

Bravais and Bravais disagreed. The ratio of two consecutive Fibonacci numbers tends to the golden ratio and the divergence angle between successive leaves in a bud, and also between successive floral organs in spiral flowers, is the *irrational golden angle*: therefore, there is no exact overlap of one organ with another underneath. Schimper and Braun's theory therefore appears to be based on imprecise measurements.

The continued fraction of an irrational number which is the solution of a quadratic equation with rational coefficients is periodic above a certain rank (Lagrange 1770 – who was a great mathematician and astronomer, first professor of analysis at the École Polytechnique upon its opening in 1794).

Auguste Bravais knew the properties of continued fractions: the convergence toward the golden ratio oscillates around this number. If one term of the sequence is greater than the golden ratio, then the next one will be smaller, and the next one greater, and so on, each being closer to the golden ratio than its predecessor (Figure 3.7).

Thus, the Bravais brothers (1830) observed that the sixth sepal (K6) of the tulip tree (*Liriodendron*, which was introduced in France in the 18th century) deviates to the left of the vertical axis from the first sepal (K1) by a non-negligible angle and the ninth sepal (K9) to the right of this axis. Then we can count eight and 13 spirals of stamens and five and eight spirals of carpels, which is characteristic of a spiral arrangement (Figure 6.13).

Box C.1. *Periodic continued fractions*

The giant of 19th-century botany, August Eichler, profusely used the notation of the ratio of two integers (e.g. 2/5) without even giving its definition; he who was otherwise so rigorous. Thus, *he saw many whorls and few spirals* ("*Die Blüthenphyllome sind meist in Quirlen, seltener spiralig angeordnet*" (Eichler 1875, p. 8). Our contemporaries are still doing this.

However, the transition from one whorl to the next is spiral. Of course, because the cycles alternate, the spiral is not continuous, as it would be in truly spiral flowers, but there is large step from the last organ of one whorl to the first one of the next whorl. This forms the basis of the prosenthesis theory formulated by Schimper and Braun (1835) and precursor to pseudo-whorls.

According to Schimper and Braun, some whorled cycles may have been initiated in a spiral. Thus, in quincuncial flowers, the alternation between the sepal cycle and the petal cycle was interpreted as a sign of the spiral initiation of organs that become whorled at anthesis. It is therefore necessary to distinguish the cycle – Cyklus – from the whorl – Quirl.

Eichler (1875) replied that in many cases, sepals are initiated successively, whereas petals are initiated simultaneously. However, he did recognize that in Ternstroemiaceae, for example, the two cycles are quincuncial.

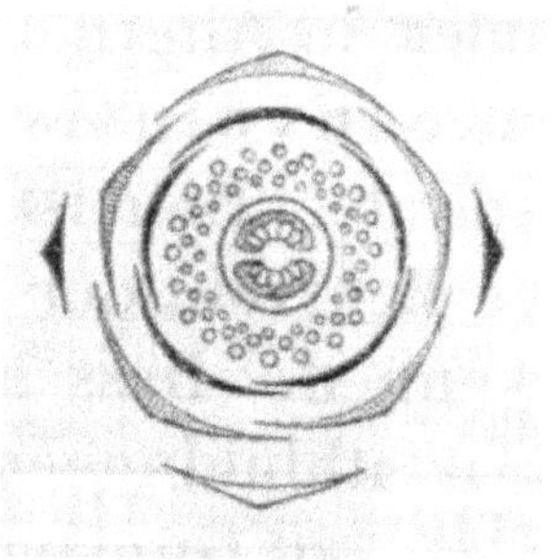

Figure C.1. *Floral diagram of* Ternstroemia silvestris *(Eichler 1878)*

For Schimper and Braun, whorls, whether at the vegetative or reproductive stage, could be interpreted as "flattened" – zusammengezogene – spirals. Eichler retorted that already in ancestral plants like the Characeae (the Charales appeared during the Silurian), one can observe pure whorls. "2/5" calyxes are frequent in plants with whorled leaves (opposite-decussate) as described in Chapter 7, for example, in *Salvia verticillata*. It was therefore more natural for Eichler to suppose that these calyces are also initiated in a whorl, rather than undergo an abrupt transition from whorled to spiral (Eichler 1875, pp. 13–18).

By contrast, Bravais and Bravais (1830) underlined the frequency of the transition from a decussate to a spiral system, as observed in most Caryophyllaceae, Hypericaceae and Cistaceae, the pentamerous calyces of which are spiral. Indeed, it is easy to observe that the fifth sepal of the terminal flower "is below the vertical at an angle suitable for the spiral system". In addition, the spiral racemes of *Verbena officinalis* (Lamiales), *Circaea lutetiana* (Rosids) and *Antirrhinum majus* (Scrophulariales) develop from decussate stems.

The Bravais brothers (1837b) studied the passage from a decussate arrangement to a spiral arrangement, which is frequent during the transition from the vegetative stage to the reproductive stage (*Salicaria*, Dipsacaceae, Asteraceae, *Antirrhinum*, *Verbena*). They measured the divergence angle in the case of *Spartium junceum* (Fabaceae; Table C.1).

Box C.2. *Transition from vegetative to reproductive stage*

Organ	Divergence angle (°)
1	90
2	170
3	114
4	143
5	140
6	120
7	155
8	131
9	138
Mean angle = 134°	

Table C.1. *Divergence angle during the transition between vegetative and reproductive stage in Spartium junceum (Bravais and Bravais 1837b). Number 0 is the bract*

C.3. Macroevolution

Bonnet's (1754) view of phyllotaxis could be considered "finalist": "transpiration of the leaves requires that air circulates freely around them and that they overlap as little as possible".

According to Godin et al. (2020), "natural selection is not on any number of spirals or divergence angle, but on the choice between Fibonacci spirals and quasi-symmetry. Before invoking natural selection to explain a shape, this suggests that one has first to consider the few possibilities left downstream of developmental canalization".

This approach seems to underestimate the fact that structure and function do coevolve. The noble angles of spiral phyllotaxis, the golden angle being only one of these, optimize the packing of organs within a meristem but also light reception by the leaves (Chapter 3).

We have seen macroevolutionary processes in action in Chapter 6. Genetically, these orders differ from basal angiosperms, as shown from their position on the angiosperm phylogenetic tree. Flowers in basal angiosperms are either strictly whorled or have a Fibonacci phyllotaxis, with the exception of Nymphaeaceae whose numerous stamens have rendered their flowers quasi-symmetric.

Strictly whorled phyllotaxis are under strong developmental constraints, especially in terms of symmetry. Even a slight perturbation affecting this symmetry is enough to cause a shift from whorled to spiral phyllotaxis (e.g. the vegetative development of *Linum* starting from the two cotyledons; the experiment of Snow and Snow). Whorled phyllotaxis are therefore incapable of macroevolution.

The possibility to diversify and thus to evolve must therefore be searched among spiral phyllotaxis. The advantage of a whorled arrangement is its coherence between phyllotaxis and organ differentiation; the advantage of a spiral arrangement is more compact packing. The fronts of Fibonacci spirals form quasi-whorls with, depending on the plastochron ratio, three, five, eight, etc. organs located on the same crown around the meristem apex. This allows the formation of pseudo-whorls. Pseudo-whorled phyllotaxis differs from spiral phyllotaxis by having a larger plastochron ratio between two successive fronts than within each pseudo-whorl (Figure C.2).

Magnoliales thus have perianths made up of multiple trimerous pseudo-whorls, followed by numerous stamens, the discontinuity in the plastochron ratio being responsible for the irregular spirals of the androecium. In *Liriodendron*, we can see one sepaloid and two petaloid pseudo-whorls, all of which are trimerous. Pseudo-whorls are made up of five organs in *Doryphora aromatica* and eight organs in *Daphnandra repandula* (Laurales).

There are around 300,000 species of angiosperms, of which 3,000 belong to Magnoliales and 3,000 to Laurales (by comparison there are 6,000 species of mammals). Within these orders, there are much fewer "missing links" in the evolutionary process than in the animal kingdom, allowing us to have a good understanding of the evolutionary mechanisms involved.

The evolution of pseudo-whorls marks the beginning of a coherence between spiral phyllotaxis and organ differentiation (synorganization; see the discussion in Chapter 6). The number of tepals, which is highly variable in basal angiosperms, thus becomes more stable.

C.4. The ancestral flower

Sauquet et al. (2017) assembled a database of floral traits (PROTEUS) from 792 species from 63 orders (98% of all plant orders) and 372 families (86% of all plant families). They constructed chronograms from molecular data calibrated with 136 fossils and provided a model-based reconstruction of the "ancestral flower" using different methods of phylogenetic analysis (maximum parsimony, maximum likelihood and a reversible-jump Markov Chain Monte Carlo Bayesian approach that allows the simultaneous exploration of multiple models of morphological evolution).

From their study, the ancestral flower was inferred to be bisexual and radially symmetric, with more than two *whorls* of three perianth organs (undifferentiated tepals), more than two *whorls* of three stamens and five spirally arranged carpels. Their analysis suggested that the perianth was *radially symmetric* (actinomorphic), stamens had introrse anthers (pollen is shed toward the center of the flower), carpels were most likely spirally arranged and all floral organs were free (not fused).

The authors concluded that the perianth and androecium of the ancestral flower in angiosperms were most likely organized in whorls rather than in a spiral, going against the prevailing idea among botanists that whorled phyllotaxis of floral organs evolved from spiral phyllotaxis.

These results were criticized by Sokoloff et al. (2018). They used the data from PROTEUS to calculate the occurrence of different combinations of whorled, spiral and uncertain phyllotaxis of the perianth and the androecium (Table C.2). They insisted that there was an absence of convincing examples of transitions from a whorled androecium to a spiral gynoecium.

Perianth	Androecium			
	Whorled	**Spiral**	**Uncertain**	**Total**
Whorled	435	3	241	679
Spiral	2	15	7	24
Uncertain	14	3	72	89
Total	451	21	320	792

Table C.2. *Occurrence of different combinaison of phyllotaxic patterns in androecium and perianth of angiosperm flowers based on PROTEUS according to Sokoloff et al. (2018)*

The proportion of whorled perianths in the database (96% of non-uncertain) suggests that pseudo-whorls have been scored as whorled. Among the sampled species coded for perianth symmetry, 78% were actinomorphic (polysymmetric), 17% were zygomorphic (bilaterally symmetrical), 3% spiral (including *Peumus boldus*) and 2% were asymmetric or disymmetric.

One major problem is that phyllotaxic patterns can differ between early (the organ initiation stage) and late flower development. Sokoloff et al. (2018) cited the example of *Illicium* where all floral organs are initiated on a spiral but the carpel is whorled at anthesis. Contact pressure acting at the post-meristematic developmental phase often transforms spirals into whorls (see Chapter 5). Thus, in PROTEUS the phyllotaxis of the gynoecium of *Illicium* is scored as ambiguous.

Sauquet et al. (2018) replied that the lack of developmental data for most species made it difficult or impossible to score traits, especially phyllotaxis, during development. Indeed, scoring some species at the early developmental stage and others at the anthetic stage produces an inconsistent database. Alternatively, scoring phyllotaxis only in species for which good developmental data are available would have led to an even greater proportion of missing data with a probably biased distribution.

Our approach is based only on developmental data and explains why many flowers have a pseudo-whorled arrangement even though they seem whorled at anthesis.

Sauquet et al. (2018) admitted that even when both data from early developmental stages and anthesis are included, there is a lot of missing data. In the case of Magnoliales and Laurales, the family level (on average PROTEUS has two species per family) is too broad to cover the diversity of phyllotaxic patterns in these orders, this diversity being a mark of great flexibility. All the meristems we modeled have different plastochron ratios. In Laurales, the flexibility of phyllotaxis is most evident in certain unisexual species where the male flower is spiral and the female flower is whorled (*Levieria acuminata*) or vice versa (*Hedycarya angustifolia*). It would be pointless to try to analyze this using cladistics. In general, the radial and angular location of the first primordia (setting the initial conditions of the model) and the plastochron ratios between floral organs differ from one species to the next. The automaton producing a phyllotaxic pattern is complex and angiosperms have inherited this mechanism with its tremendous variability. The possibility of spiral and whorled phyllotaxis has probably existed since the very beginning of angiosperm evolution, which may explain the presence of both in basal angiosperms.

For Sokoloff et al. (2018), it is difficult to accept that the ancestral flower featured a combination of two character states, namely a whorled androecium and a spiral gynoecium, which is unknown in any extant species. They suggest that the ancestral flower was either entirely whorled or entirely spiral.

The reconstructed ancestral flower of Sauquet et al. (2017) differs significantly from that proposed by Arber and Parkin (1907), which looked like a *Magnolia* flower, a widely held view among botanists over the past century. *Magnolia* has a whorled perianth and a spiral androecium.

According to our analysis, the ancestral flower of Sauquet et al. looks like an ancestor of *Daphnandra repandula* (Figure 6.19; the latter does not feature in PROTEUS). *D. repandula* has a pseudo-whorled perianth and androecium and a spiral gynoecium. The ancestral flower of Sauquet et al. has an undifferentiated perianth, which reinforces the idea that it is not whorled but pseudo-whorled. The perianth of *D. repandula* is octomerous; the diameter of the front of the floral meristem is reduced to allow this high merosity while retaining a viable flower size. On the contrary, the ancestral flower of Sauquet et al. could have a high plastochron ratio in its perianth and therefore a trimerous meristem front like Magnoliaceae.

Primordium number	Plastochron ratio	Primordium number	Plastochron ratio
K1	1.19	S2	1.07
K1	1.17	S2	1.06
K1	1.30	S2	1.30
K2	1.13	S3	1.06
K2	1.12	S3	1.06
K2	1.30	S3	1.30
K3	1.10	C	1.06
K3	1.09	C	1.05
K3	1.30	C	1.05
S1	1.08	C	1.05
S1	1.07	C	1.05
S1	1.30	C	1.05

Table C.3. *Plastochron ratios used to model the ancestral flower of Sauquet et al. (2017)*

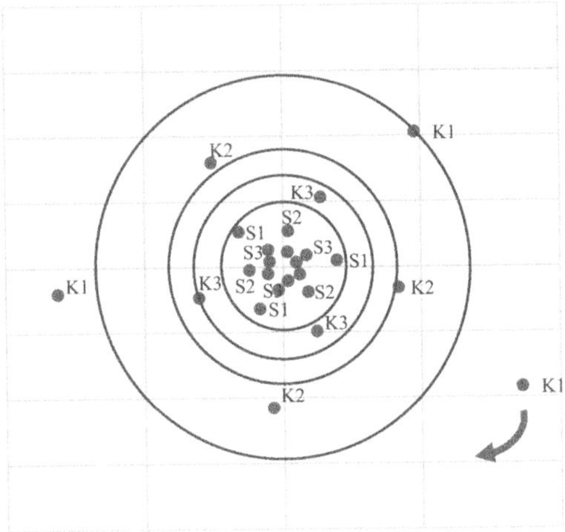

Figure C.2. *Modeling of the ancestral flower of Sauquet et al.; the centric representation shows pseudo-whorls in the perianth and androecium and spirally arranged carpels (unlabeled dots in the center)*

We chose the parameters of a Fibonacci spiral (as in *Illicium*) starting from the fifth organ after the vegetative spiral (this spiral continues into the floral meristem, as in *Magnolia*) and imposed pseudo-whorls in the perianth and androecium, as in some Laurales (Table C.3; Figure C.2). The gynoecium is spiral. Unlike *D. repandula*, the diameter of the front is considered equal to 1 as in basal angiosperms and Magnoliales, which corresponds to flowers with a low perianth merosity (= 3).

C.5. Inhibition versus disk-stacking

Using a single inhibition threshold for all flowers, Nakagawa et al. (2020) were able to model a wide range of perianth types. However, by only taking two sources of inhibition into account, that is, an adaxial and an abaxial inhibition source (and thus with bilateral symmetry), they restrict themselves to flowers with bilateral symmetry (in accordance with Curie's principle). In addition, even among bilaterally symmetrical flowers, there are many cases (such as

Antirrhinum) where we feel it is difficult to find only one dorsal source of inhibition.

Douady and Couder's model, where the only parameter is the inhibition threshold, rapidly showed its limits (Douady and Couder 1996b; Smith et al. 2006; Kitazawa and Fujimoto 2015). This type of model only allows one to model, with a few exceptions, very generic flowers.

As an alternative, research focused on the disk-stacking method, which allows a continuous decrease in disk diameter (the equivalent of the plastochron ratio) from larger to smaller disks (Zagórska-Marek and Szpak 2008; Atela 2011; Douady and Golé 2016). This approach is more suitable to model *cylindrical lattices* such as those of the *Plantago* inflorescence, the generative shoot and gynoecium of *Magnolia* (Zagórska-Marek and Szpak 2008), the vegetative stem of cabbage, the artichoke inflorescence, the birch catkin, the cones of *Pinus nigra* and *Cedrus libani*, which form large lattices of plant organs (Douady and Golé 2016). The disk-stacking method also allows one to model rising phyllotaxis (traveling down the van Iterson diagram). This method, although it does not consider the molecular and cellular mechanisms that control phyllotaxis, explains in a simple way how primordia produce the patterns commonly found in plants.

By contrast, our approach models *real centric lattices*, which allows us to use data from floral diagrams (from Eichler 1875, 1878) and SEM images. Furthermore, our model is parametrized by the plastochron ratio, which is a measurable physical character.

Both approaches give fundamentally equivalent results in terms of primordia initiation at the front, rising phyllotaxis to noble angles (the van Iterson diagram or our phyllotaxic frieze), organ permutations and quasi-symmetric arrangements (Golé et al. 2016; Godin et al. 2020; our results in Chapter 6).

However, the physical basis of the two models is very different. The disk-stacking model relies on contact pressure between primordia, represented as incompressible disks, and pressure exerted by a "wall" on these organs. The rise in phyllotaxis mode is determined by the rate of decrease in organ size relative to the size of the center of the meristem and by competition between primordia for space. Development is thus channeled in both the literal (by the walls) and figurative sense: phyllotaxis is primarily governed by geometric developmental constraints (Godin et al. 2020). However, according to our model, the interaction between organs is not just a purely mechanical force exerted by hard disks in direct contact; the inhibition field plays an important organizational role even when two primordia are at "relatively important distance" from each other.

The mechanical and biochemical approaches yield the same results. The selective success of the physicochemical processes underlying plant growth is due to the fact that these processes optimize compaction during development, as shown by the disk-stacking model (see Chapter 3).

Newell et al. (2008) have shown that the partial differential equation governing the biochemical process is almost identical in form to that governing a mechanical (buckling) model, and indeed to that governing the behavior of a universal class of pattern-forming systems. They suggest that both biochemical processes (auxin localization) and mechanical forces (non-uniform stress distribution) may together play key roles in phyllotaxis.

C.6. Inflorescences and contact pressure

According to Church (1904), contact pressure applies to primordia only after they are formed. Contact pressure may thus lead to secondary displacements from the site of initiation and may come from three sources:

– older primordia exert pressure on younger primordia after they are initiated;

– growing primordia exert reciprocal pressure when there is direct contact;

– pressure can be the effect of a rigid boundary on a growing system.

The architecture of cymose inflorescences is based on the initiation of bracteoles that subtend the floral meristems. Only a few species with strictly racemose inflorescences have bracteoles (e.g. *Delphinium*, some Fabaceae). In the latter, the bracteoles influence sepal arrangement but are sterile: no flower develops in their axil. According to Bull-Hereñu et al. (2022), the sequence of organ initiation in many monosymmetric (bilaterally symmetrical) flowers tends to be unidirectional, often from the ventral to the median then the dorsal side (e.g. many Fabaceae, Lamiales), and this is probably related to the fact that the dorsal side exerts the strongest pressure. Indeed, we have shown that in *Baptisia australis* (Papilioinoideae) raceme the two upper bracts above a developing flower are necessary to reinforce the inhibition produced by the two bracteoles of this flower and produce a unidirectional floral organ initiation pattern (Figure 5.11(C)).

In the absence of bracteoles, Eichler suggested that the position of the first sepals of a flower is a consequence of the position of the last leaves. This is the case for *Geranium robertianum* (Figure 7.11), but there are other scenarios.

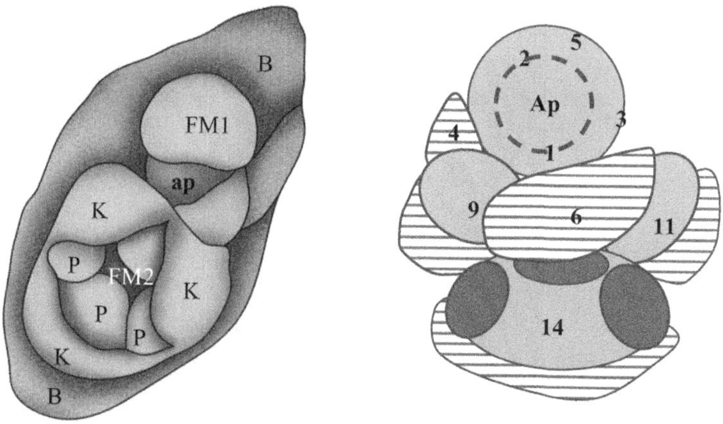

Figure C.3. *(A) Partial inflorescence of Telipogon falcatus (from Pabón-Mora and Gonzalez 2008). ap = apex. (B) Diagram of part of the racemose inflorescence of Antirrhinum majus (from Carpenter et al. 1995): numbers correspond to the nodes*

Telipogon belongs to Orchidaceae (monocots): most species of this genus have the typical monosymmetric orchid perianth. The inflorescence of *T. falcatus* is racemose: in the absence of bracteoles, the source of dorsal inhibition is the apex of the inflorescence (Figure C.3(A)).

In many plants, such as *Antirrhinum majus*, with pentamerous flowers in racemose inflorescences, the calyx starts to develop only after the initiation of numerous meristems (Figures C.3(B) and 7.8(A)). We can consider that the field shaping the perianth is not due to the biochemical activity of the neighboring meristems or bracts (reserved for central primordia: Figure 5.4(B)), but to contact pressure. Already for Schwendener (1878), the two nearest primordia located on the two parastichies crossing a given primordium *i* produce, by contact, a pressure on this primordium *i*. Bull-Hereñu et al. (2022) have noted that when the bracts in racemose inflorescences are "rigid", they reduce the available space for axillary meristems and compress young developing flowers. When the pressure exerted by the bracts is high, the floral primordium becomes squeezed and even flattened (see FM14 on Figure C.3(B)).

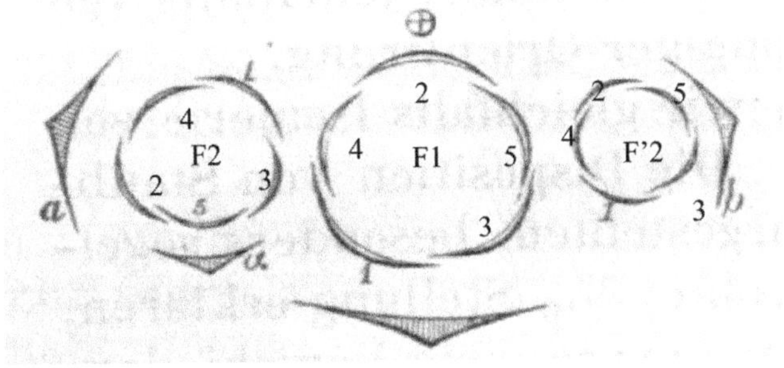

Figure C.4. *Diagram of a partial inflorescence of Caltha palustris (from Eicher 1878)*

Bull-Hereñu et al. (2022) have noted that in inflorescences with a terminal flower, the terminal flower is not squeezed by a subtending bract as it develops directly from the transformation of the inflorescence meristem. The lateral flowers, meanwhile, present an altered phyllotaxis. We have observed that this was the case for lateral flowers of *Caltha palustris*, and that to model them, an adaxial source of "repulsion" is required for F2 and F2' (Figure C.4; Chapter 7). These sources have to be located near the point of contact between the lateral flower and the terminal flower, so we can consider that the repulsion field generated by F1 is due to contact pressure.

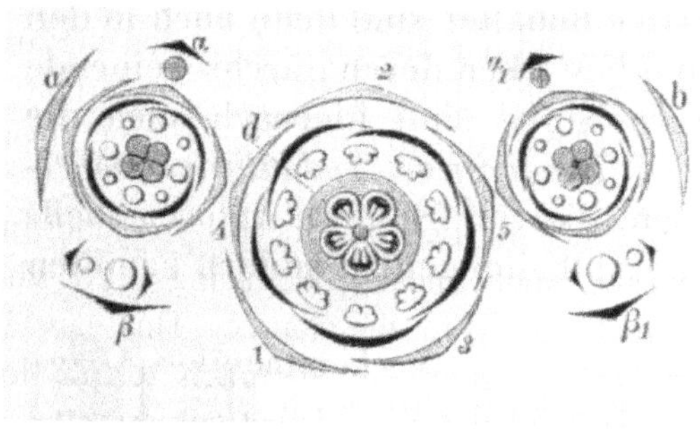

Figure C.5. *Diagram of a partial inflorescence of Ruta graveolens (Rutaceae) (from Eicher 1878). Note that a and b are the two last cauline leaves*

According to Bull-Hereñu et al. (2022), there is also a difference in merosity between lateral and terminal flowers in *Trochodendron*, *Ruta* and *Adoxa* as a result of the compression exerted by bracts and the "inflorescence axis" on lateral floral meristems. The terminal flower of *Ruta graveolens* has a 5A type quincuncial arrangement, due to the effect of the two leaves, a and b, whose divergence angle is close to 137°. The initiation of the tetramerous perianth of lateral flowers is the consequence of four inhibition/repulsion sources that are opposite-decussate (Walch and Blaise 2022b). They are affected by inhibition from their two bracteoles as well as pressure coming from both axial sides, where the "jaws" are their subtending bract and the lateral sepals of the terminal flower. However, the fact that the two outer sepals of lateral flowers are on the abaxial/adaxial axis proves that the intensity of this pressure is less than the lateral inhibition (Figure C.5).

In *Salicornia europeae*, one can think that the influence of the fused margins of the involucral bracts, which makes the axes of flowers F2 and F2' converge toward that of F1, is due to contact pressure rather than a biochemical factor. In *Frankenia pulverulenta*, the initiation of the secondary involucra requires the contact pressure exerted laterally by a primary involucral leaf, and by the terminal flower. In *Bougainvillea spectabilis* and *Anemone ranunculoides*, the phyllotaxis of lateral flowers is determined by two axial sources of repulsion: an abaxial involucral leaf and the adaxial central flower. They exert their influence by contact pressure. When the lateral flowers of *Mirabilis violacea* develop, the pentamerous involucrum is already fully formed and has already determined the arrangement of the sepals in the terminal flower. Its very important influence on lateral flowers may also be due to contact pressure (Chapter 7).

If organs are "soft", the field produced by contact pressure could be modeled as in the biochemical model using a law where the effect decreases with distance from the source. This law was used by Levitov (1991b) in his energetic approach to phyllotaxis, which models "soft lattices" subjected to strong compression. The network of primordia is "soft", that is, it is not composed of rigid disks as in the disk-stacking model. This pressure field (and not inhibition field) was also used by Kitazawa and Fujimoto (2015) to model the post-meristematic development of floral meristems. We successfully used it to model the post-meristematic development of *Illicium* (Chapter 5). Thus, *our model integrates both biochemical inhibition and contact pressure* between soft organs.

Contact pressure models (Chapter 2) show that in a *2D* lattice, contact pressure that minimizes the distance between organs (Adler 1977) or results in primordia rearrangement (Ridley 1982) produces Fibonacci spirals. In a *1D* circular model, it leads to a pseudo-whorled arrangement (see *Illicium*; Figure 5.10).

This complementarity between biochemical field and mechanical force not only determines the position of organs but also how these fit together perfectly, without any "holes" between them. Thus, the partial *cincinnus* of *Phenakospermum guyanense* (Strelitziaceae) (Figure C.6(A)) gives the impression of having been built from modeling clay, the archetype of a malleable material. The extreme compaction that results from it is also particularly visible on the thyrse of *Salvia verticillata* (Figure 7.7).

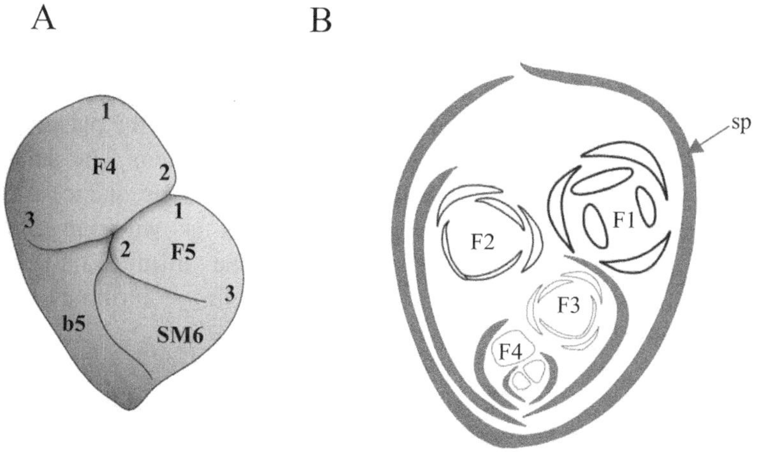

Figure C.6. *Phenakospermum guyannese (redrawn from Kirchoff 2003). (A) The apical region of the cincinnus with the sympodial meristem of the sixth flower (SM6), a bracteole (b5), two floral meristems (F4 and F5) and the position of their sepals (1,2,3). (B) Diagram of the cincinnus. sp = spathe*

Unlike in Levitov's model or the disk-stacking model, contact pressure acting on an inflorescence may not be due to the surrounding environment. This can be seen in racemose inflorescences (e.g. *Antirrhinum majus*). The developmental constraint described by the disk-stacking model is therefore not an external constraint but is generated within the inflorescence itself like in prestressed concrete. The same applies to vegetative buds of annual plants that lack bud scales (e.g. *Linum usitatissimum*), where contact pressure would be a consequence of growth itself.

By contrast, in *Salvia verticillata* (Figure 7.23), we can see a large bract (which needed to be removed to reveal the floral meristems) squeezing the cyme harboring two *cincinni* fitting together like the two sides of a zipper (fitting all the better as these *cincinni* are antidromous). Similarly, the floral meristem of *Phenakospermum guyannese* is subjected to pressure from the spathe surrounding the entire inflorescence (Figure C.6(B)).

Phyllotaxy thus appears to be determined in a general way by a biochemical inhibition field that controls the development of the different vegetative and reproductive organs and by contact pressure between these organs as they grow in size. The field decreases with the distance between each organ so that this distance (and thus organ size since they are next to each other), as measured by the plastochron ratio, determines organ arrangement within a meristem. The detailed arrangement of organs in each flower is thus the consequence of the sequence of plastochron ratios. Each species has its own arrangement: this sequence is thus *genetically* determined. The "infinite" number of possible combinations of these plastochron ratios contributes to the immense diversity of plants.

Levitov (1991b) used a powerful mathematical tool (group theory) to show that very different phenomena such as phyllotaxis, Rayleigh-Bénard convection (the movement of a liquid in a container with a temperature gradient under the influence of the Earth's rotation) (Rivier et al. 1984) or flux lattices in layered superconductors, to which we can add drops of ferrofluid in a magnetic field (the Douady and Couder experiment), can produce Fibonacci spirals. Although the "engines" of these phenomena are very different (auxin concentration, energy transfer or a magnetic field), they all obey the same law of symmetry, in the broad sense that this notion has acquired in modern physics. This explains the numerous geometrical analogies in fields that are very different from each other (D'Arcy Thompson 1917; Weyl 1952).

The phyllotaxic engine is based on a principle of optimization (minimizing the inhibition potential) that gives specific geometrical properties to plants such as Fibonacci spirals and ensures optimal packing, as illustrated by the disk-stacking model. However, primordia are not rigid structures but soft organs, subjected to contact pressure that ensures their contiguity. Taken together, these properties, as well as the rich parametrization of the phyllotaxic engine lie behind its formidable evolutionary power.

References

Adler, I. (1974). A model of contact pressure in phyllotaxis. *Journal of Theoretical Biology*, 45, 1–79.

Adler, I. (1977). The consequences of contact pressure in phyllotaxis. *Journal of Theoretical Biology*, 65, 29–77.

Adler, I., Barabé D., Jean R. (1997). A history of the study of phyllotaxis. *Annals of Botany*, 80, 231–244.

Alvarez-Buylla, E., Benítez, M., Corvera-Poiré, A., Cador, A.C., de Folter, S., de Buen, A.G., Garay-Arroyo, A., Garcia-Ponce, B., Jaimes-Miranda, F., Pérez-Ruiz, R.V. et al. (2010). Flower development. *The Arabidopsis Book*, 2010(8). American Society of Plant Biologists, Rockville.

APG IV (2016). An update of the Angiosperm Phylogeny Group classification for the orders and families of flowering plants. *Botanical Journal of the Linnean Society*, 181, 1–20.

Arber, E.A.N. and Parkin, J. (1907). On the origin of angiosperms. *Journal of the Linnean Society of London*, 38, 29–80.

Atela, P. (2011). The geometric and dynamic essence of phyllotaxis. *Mathematical Modelling of Natural Phenomena*, 6, 173–186.

Bachelard, G. (1938). *La formation de l'esprit scientifique*. Vrin, Paris.

Blaise, S. (2011). Architecture inflorescentielle de quelques Myosotis d'Europe occidentale. *Journal de Botanique de la Société Botanique de France*, 53, 21–30.

Bonnet, C. (1754). *Recherches sur l'usage des feuilles dans les plantes*. Elie Luzac fils, Göttingen/Leiden.

Bonnet, C. (1764). *Contemplation de la nature*. Marc-Michel Rey, Amsterdam.

Braun, A. (1835). Dr. Carl Schimper's Vorträge über die Möglichkeit eines wissen-schaftlichen Verstandnis der Blattstellung, nebst Andeutung de Hauptsächlichen Blattstellungsgesetze und Insbesondere der Neuentdeckten Gesetze der Aneinanderreihung von Cyclen Verschiedener Maasse. *Flora*, 18, 145–191.

Bravais, A. (1850). Mémoire sur les systèmes formés par les points distribués régulièrement sur un plan ou dans l'espace. *Journal de l'École Polytechnique*, 19, 1–128.

Bravais, L. and Bravais, A. (1830). Essai sur la disposition générale des feuilles rectirisées. In *Congrès scientifique de France*. Imprimerie de Perol, Clermont-Ferrand.

Bravais, L. and Bravais, A. (1837a). Essai sur la disposition des feuilles curvisériées. *Annales des Sciences Naturelles Botanique*, 7, 42–110.

Bravais, L. and Bravais, A. (1837b). Essai sur la disposition symétrique des inflorescences. *Annales des Sciences Naturelles Botanique*, 7, 193–221, 291–348.

Bull-Hereñu, K., dos Santos, P., Toni, J.F.G., El Ottra, J.H.L., Thaowetsuwan, P., Jeiter, J., Ronse de Craene, L.P., Iwamoto, A. (2022). Mechanical forces in floral development. *Plants*, 11, 661.

de Candolle, A.-P. (1813). *Théorie élémentaire de botanique*. Déterville, Paris.

de Candolle, A.-P. (1827). *Organographie végétale*. Germer Baillière, Paris.

Carpenter, R., Copsey, L., Vincent, C., Doyle, S., Magrath, R., Coen, E. (1995). Control of flower development and phyllotaxy by meristem ldentity genes in antirrhinum. *The Plant Cell*, 7, 2001–2011.

Castel, R., Kusters, E., Koes, R. (2010). Inflorescence development in *Petunia*: Through the maze of botanical terminology. *Journal of Experimental Botany*, 61, 2235–2246.

Chadefaud, M. and Emberger, L. (1960). *Traité de Botanique systématique*. Masson, Paris.

Chandler, J.W. and Werr, W. (2014). *Arabidopsis* floral phytomer development: Auxin response relative to biphasic modes of organ initiation. *Journal of Experimental Botany*, 65, 3097–3110.

Chandler, J.W., Jacobs, B., Cole, M., Comelli, P., Werr, W. (2011). DORNRÖSCHEN-LIKE expression marks Arabidopsis floral organ founder cells and precedes auxin response maxima. *Plant Molecular Biology*, 76, 171–185.

Chase, M.W., Soltis, D.E., Olmstead, R.G., Morgan, D., Les, D.H., Mishler, B.D., Duvall, M.R., Price, R.A., Hills, H.G., Qiu, Y.-L. et al. (1993). Phylogenetics of seed plants: An analysis of nucleotide sequences from the plastid gene rbcL. *Annals of the Missouri Botanical Garden*, 80, 528–548, 550–580.

Church, A.H. (1904). *On the Relation of Phyllotaxis to Mechanical Laws*. Williams and Norgate, London.

Claßen-Bockhoff, R. and Bull-Hereñu, K. (2013). Towards an ontogenetic understanding of inflorescence diversity. *Annals of Botany*, 112, 1523–1542.

Couder, Y. (1998). Initial transitions, order and disorder in phyllotactic patterns: The ontogeny of *Helianthus annuus*. A case study. *Acta Societatis Botanicorum Poloniae*, 67, 129–150.

Curie, P. (1894). Sur la symétrie dans les phénomènes physiques, symétrie d'un champ électrique et d'un champ magnétique. *Journal de physique théorique et appliquée*, 3, 393–417.

D'Arcy Thompson, W. (1917). *On Growth and Form*. Cambridge University Press, Cambridge.

Deschatres, R. (1954). Recherches sur la phyllotaxie du genre *Sedum*. *Revue générale de botanique*, 61, 501–570.

Douady, S. and Couder, Y. (1992). Phyllotaxis as a physical self-organized growth process. *Physical Review Letters*, 68, 2098–2101.

Douady, S. and Couder, Y. (1996a). Phyllotaxis as a dynamical self organizing process. Part I: The spiral modes resulting from time-periodic iterations. *Journal of Theoretical Biology*, 178, 255–274.

Douady, S. and Couder, Y. (1996b). Phyllotaxis as a dynamical self-organizing process. Part II: The spontaneous formation of a periodicity and the coexistence of spiral and whorled patterns. *Journal of Theoretical Biology*, 178, 275–294.

Douady, S. and Couder, Y. (1996c). Phyllotaxis as a dynamical self-organizing process. Part III: The simulation of the transient regimes of ontogeny. *Journal of Theoretical Biology*, 178, 295–312.

Douady, S. and Golé, C. (2016). Fibonacci or quasi-symmetric phyllotaxis. Part II: Botanical observations. *Acta Societatis Botanicorum Poloniae*, 85, 3534.

Eichler, A.W. (1875). *Blüthendiagramme construirt und erläutert*, Volume I. Engelmann, Leipzig.

Eichler, A.W. (1878). *Blüthendiagramme construirt und erläutert*, Volume II. Engelmann, Leipzig.

El, E., Remizowa, M., Sokoloff, D. (2020). Developmental flower and rhizome morphology in *Nuphar* (Nymphaeales): An interplay of chaos and stability. *Frontiers in Cell and Developmental Biology*, 8, 303.

Endress, P. (2001). The flowers in extant basal angiosperms and inferences on ancestral flowers. *International Journal of Plant Sciences*, 162, 1111–1140.

Endress, P. (2011). Evolutionary diversification of the flowers in angiosperms. *American Journal of Botany*, 98, 370–396.

Endress, P. (2016). Development and evolution of extreme synorganization in angiosperm flowers and diversity: A comparison of Apocynaceae and Orchidaceae. *Annals of Botany*, 117, 749–767.

Endress, P. and Armstrong, J. (2011). Floral development and floral phyllotaxis in *Anaxagorea* (Annonaceae). *Annals of Botany*, 108, 835–845.

Endress, P. and Doyle, J. (2015). Ancestral traits and specializations in the flowers of the basal grade of living angiosperms. *Taxon*, 64, 1093–1116.

Erbar, C. and Leins, P. (1994). Flowers in Magnoliidae and the origin of flowers in other subclasses of the angiosperms. Part I: The relationships between flowers of Magnoliidae and Alismatidae. *Plant Systematics and Evolution*, S8, 193–208.

Erickson, R. (1973). Tubular packing of spheres in biological fine structures. *Science*, 181, 705–716.

Galv. n-Ampudia, C., Cerutti, G., Legrand, J., Brunoud, G., Martin-Arevalillo, R., Azai, R., Bayle, V., Moussu, S., Wenzl, C., Jaillais, Y. et al. (2020). Temporal integration of auxin information for the regulation of patterning. *eLife*, 9, e55832.

Godin, C., Golé, C., Douady, S. (2020). Phyllotaxis as geometric canalization during plant development. *Development*, 147, dev165878.

Golé, C., Dumais, J., Douady, S. (2016). Fibonacci or quasi-symmetric phyllotaxis. Part I: Why? *Acta Societatis Botanicorum Poloniae*, 85, 3533.

Green, P.B. (1999). Expression of pattern in plants: Combining molecular and calculus-based paradigms. *American Journal of Botany*, 86, 1059–1076.

Guinochet, M. (1965). *Notions fondamentales de botanique générale*. Masson, Paris.

Hellwig, H., Engelmann, R., Deussen, O. (2006). Contact pressure models for spiral phyllotaxis and their computer simulation. *Journal of Theoretical Biology*, 240, 489–500.

Hepworth, S.R., Zhang, Y., McKim, S., Li, X., Haughn, G.W. (2005). BLADE-ON-PETIOLE-dependent signaling controls leaf and floral patterning in Arabidopsis. *Plant Cell*, 17, 1434–1448.

Hofmeister, W. (1868). *Allgemeine Morphologie der Gewachse*. Engelmann, Leipzig.

Hotton, S., Johnson, V., Wilbarger, J., Zwieniecki, K., Atela, P., Golé, C., Dumais, J. (2006). The possible and the actual in phyllotaxis: Bridging the gap between empirical observations and iterative models. *Journal of Plant Growth Regulation*, 25, 313–323.

Hrabovsky, M. (2019). A rhipidium is not a cyme. *Acta Botanica Comentianae*, 54, 57–59.

van Iterson, G. (1907). *Mathematische und Mikroskopisch-Anatomische Studien über Blattstellungen nebst Betraschtungen über den Schalenbau der Miliolinen*. Gustav Fischer, Jena.

Jean, R.V. (1994). *Phyllotaxis. A Systemic Study in Plant Morphogenesis*. Cambridge University Press, Cambridge.

Kitazawa, M. and Fujimoto, K. (2015). A dynamical phyllotaxis model to determine floral organ number. *PLoS Computational Biology*, 11, e1004145.

Lagrange, J.-L. (ed.) (1770). Additions au mémoire sur la résolution des équations numériques. In *Oeuvres complètes*, Volume II. Gauthier-Villars, Paris.

Levitov, L.S. (1991a). Phyllotaxis of flux lattices in layered superconductors. *Physical Review Letters*, 66, 224–227.

Levitov, L.S. (1991b). Energetic approach to phyllotaxis. *Europhysics Letters*, 14, 533–539.

Lyndon, R. (1978). Phyllotaxis and the initiation of primordia during flower development in *Silene*. *Annals of Botany*, 42, 1349–1360.

Marzec, C. and Kappraff, J. (1983). Properties of maximal spacing on a circle related to phyllotaxis and to the golden mean. *Journal of Theoretical Biology*, 103, 201–226.

Meicenheimer, R.D. (1979). Relationships between shoot growth and changing phyllotaxy of *Ranunculus*. *American Journal of Botany*, 66, 557–569.

Meicenheimer, R.D. (1987). Role of stem growth in *Linum usitatitissimum* leaf trace patterns. *American Journal of Botany*, 74, 857–867.

Meicenheimer, R.D. (2006). Stem unit growth analysis of *Linum Usitatissimum* internode development. *American Journal of Botany*, 93, 55–63.

Mitchison, G. (1977). Phyllotaxis and the Fibonacci series. *Science*, 196, 270–275.

von Nägeli, C.W. (1858). *Beiträge zur wissenschaftlichen Botanik*. Wilhelm Engelmann, Leipzig.

Nakagawa, A., Kitazawa, M.S., Fujimoto, K. (2020). A design principle for floral organ number and arrangement in flowers with bilateral symmetry. *Development*, 147, dev182907.

Newell, A.C., Shipman, P.D., Suna, Z. (2008). Phyllotaxis: Cooperation and competition between mechanical and biochemical processes. *Journal of Theoretical Biology*, 251, 421–439.

Nysten, P.-H. (1824). *Dictionnaire de médecine, de chirurgie, de pharmacie, des sciences accessoires et de l'art vétérinaire*. J.S. Chaude, Paris.

Pabon-Mora, N. and Gonzalez, F. (2008). Floral ontogeny of *Telipogon* spp. (Orchidaceae) and insights on the perianth symmetry in the family. *International Journal of Plant Sciences*, 169, 1159–1173.

Plantefol, L. (1949). *L'ontogénie de la fleur*. Masson, Paris.

Prenner, G. (2004). New aspects in floral development of Papilionoideae: Initiated but suppressed bracteoles and variable initiation of sepals. *Annals of Botany*, 93, 537–545.

Rebocho, A.B., Bliek, M., Kusters, E., Castel, R., Procissi, A., Roobeek, I., Souer, E., Koes, R. (2008). Role of EVERGREEN in the development of the cymose petunia inflorescence. *Developmental Cell*, 15, 437–447.

Remizowa, M.V., Sokoloff, D., Calvo, S., Tomasello, A. (2012). Flowers and inflorescences of the seagrass Posidonia (Posidoniaceae, Alismatales). *American Journal of Botany*, 99, 1592–1608.

Remizowa, M.V., Rudall, P.J., Choob, V.V., Sokoloff, D. (2013). Racemose inflorescences of monocots: Structural and morphogenetic interaction at the flower/inflorescence level. *Annals of Botany*, 112, 1553–1566.

Richards, F.J. (1951). Phyllotaxis: Its quantitative expression and relation to growth in the apex. *Philosophical Transactions of the Royal Society London*, B235, 509–564.

Ridley, J.N. (1982). Packing efficiency in sunflower heads. *Mathematical Biosciences*, 58, 129–139.

Rivier, N., Occelli, R., Pantaloni, J., Lissowski, A. (1984). Structure of Bénard convection cells, phyllotaxis and crystallography in cylindrical symmetry. *Journal de physique*, 45, 49–63.

Ronse de Craene, L. (2017). Floral development of *Berberidopsis beckleri* – Can an additional species of the Berberidopsidaceae add evidence to floral evolution in the core eudicots? *Annals of Botany*, 119, 599–610.

Ronse de Craene, L. (2022). *Floral Diagrams. An Aid to Understanding Flower Morphology and Evolution*. Cambridge University Press, Cambridge.

Rothen, F. and Koch, A.-J. (1989). Phyllotaxis, or the properties of spiral lattices. Part I: Shape invariance under compression. *Journal de physique*, 50, 633–657.

Rutishauser, R. (1998). Plastochron ratio and leaf arc as parameters of a quantitative phyllotaxis analysis in vascular plants. In *Symmetry in Plants*, Jean, R.V. and Barabé, D. (eds). World Scientific Publishing Co. Pte. Ltd, Singapore.

Sauquet, H., von Balthazar, M., Magallón, S., Doyle, J.A., Endress, P.K., Bailes, E.J., de Morais, E.B., Bull–Hereñu, K., Carrive, L., Chartier, M. et al. (2017). The ancestral flower of angiosperms and its early diversification. *Nature Communications*, 8, 16047.

Sauquet, H., von Balthazar, M., Doyle, J.A., Endress, P.K. (2018). Challenges and questions in reconstructing the ancestral flower of angiosperms. A reply to Sokoloff et al. *American Journal of Botany*, 105, 127–135.

Schimper, C.F. (1830). Beschreibung des *Symphytum Zeyheri* und seiner zwei deutschen Verwandten der *S. bulborum* Schimper und *S. tuberosum* Jacqu. *Geiger's Magazin für Pharmacie*, 29, 1–92.

Schoute, J.C. (1913). Beitrage zur Blattstellunglehre. Part I: Die Theorie. *Recueil de Travaux Botaniques Néerlandais*, 10, 153–339.

Schwendener, S. (1878). *Mechanische Theorie der Blattstellungen*. Engelmann, Leipzig.

Smith, R., Prusinkiewicz, P., Kuhlemeier, C. (2006). Inhibition fields for phyllotactic pattern formation: A simulation study. *Canadian Journal of Botany*, 84, 1635–1649.

Smyth, D.R., Bowman, J.L., Meyerowitz, E.M. (1990). Early flower development in *Arabídopsis*. *The Plant Cell*, 2, 755–767.

Snow, M. and Snow, R. (1935). Experiments on phyllotaxis. Part III: Diagonal splits through decussate apices. *Philosophical Transactions of the Royal Society London*, B225, 63–94.

Snow, M. and Snow, R. (1952). Minimum area and leaf determination. *Proceedings of the Royal Society of London*, B139, 545–566.

Sokoloff, D., Remizowa, M., Bateman, R., Rudall, P.J. (2018). Was the ancestral angiosperm flower whorled throughout? *American Journal of Botany*, 105, 5–15.

Souer, E., van der Krol, A.R., Kloos, D., Spelt, C., Bliek, M., Mol, J., Koes, R. (1998). Genetic control of branching pattern and floral identity during Petunia inflorescence development. *Development*, 125, 733–742.

Staedler, Y. and Endress, P. (2009). Diversity and lability of floral phyllotaxis in the pluricarpellate families of core Laurales. *International Journal of Plant Sciences*, 170, 522–550.

Staedler, Y., Weston, P., Endress, P. (2007). Floral phyllotaxis and floral architecture in Calycanthaceae (Laurales). *International Journal of Plant Sciences*, 168, 285–306.

Strauss, S., Lempe, J., Prusinkiewicz, P., Tsiantis, M., Smith, R.S. (2020). Phyllotaxis: Is the golden angle optimal for light capture? *New Phytologist*, 225, 499–510.

Tucker, S. and Hodges, S. (2005). Floral ontogeny of *Aquilegia*, *Semiaquilegia* and *Enemion* (Ranunculaceae). *International Journal of Plant Sciences*, 166, 557–674.

Veen, A. and Lindenmayer, A. (1977). Diffusion mechanism for phyllotaxis. *Plant Physiology*, 60, 127–139.

Walch, J.-P. (2022). Modeling phyllotaxis: From the inhibition potential to the real plant. *Journal of Theoretical Biology*, 553, 111261.

Walch, J.-P. and Blaise, S. (2022a). Modeling basal angiosperm floral phyllotaxis. *Botany Letters*, 169, 305–318.

Walch, J.-P. and Blaise, S. (2022b). Mathematical modeling the influence of bracteoles on sepal arrangement. *Flora*, 152138.

Weberling, F. (1989). *Morphology of Flowers and Inflorescences*. Cambridge University Press, Cambridge.

Weisse, A. (1894). Neue Beitraäge zur mechanischen Blattstellungslehre. *Jahrbücher für Wissenschaftliche Botanik*, 26, 236–294.

Weyl, H. (1952). *Symmetry*. Princeton University Press, Princeton.

Williams, R. (1975). *The Shoot Apex and Leaf Growth*. Cambridge University Press, Cambridge.

Wiss, D. and Zagórska-Marek, B. (2012). Geometric parameters of the apical meristem and the quality of phyllotactic patterns in *Magnolia* flowers. *Acta Societatis Botanicorum Poloniae*, 81, 203–216.

Zagórska-Marek, B. (1994). Phyllotaxic diversity in *Magnolia* flowers. *Acta Societatis Botanicorum Poloniae*, 63, 117–137.

Zagórska-Marek, B. and Szpak, M. (2008). Virtual phyllotaxis and real plant model cases. *Functional Plant Biology*, 35, 1025–1033.

Index

L, M, O

P, Q, R

S, T, V,

W, Z

Other titles from

iSTE

in

Biology and Biomedical Engineering

2022

DAMBRICOURT MALASSÉ Anne
Embryogeny and Phylogeny of the Human Posture 2: A New Glance at the Future of our Species (Comparative Anatomy and Posture of Animal and Human Set – Volume 4)

FOSSÉ Philippe
Structures and Functions of Retroviral RNAs: The Multiple Facets of the Retroviral Genome (Nucleic Acids Set – Volume 1)

2021

DAMBRICOURT MALASSÉ Anne
Embryogeny and Phylogeny of the Human Posture 1: A New Glance at the Future of our Species (Comparative Anatomy and Posture of Animal and Human Set – Volume 3)

GRANDCOLAS Philippe, MAUREL Marie-Christine
Systematics and the Exploration of Life

HADJOUIS Djillali
The Skull of Quadruped and Bipedal Vertebrates: Variations, Abnormalities and Joint Pathologies (Comparative Anatomy and Posture of Animal and Human Set – Volume 2)

HULLÉ Maurice, VERNON Philippe
The Terrestrial Macroinvertebrates of the Sub-Antarctic Îles Kerguelen and Île de la Possession

2020

CAZEAU Cyrille
Foot Surgery Viewed Through the Prism of Comparative Anatomy: From Normal to Useful
(Comparative Anatomy and Posture of Animal and Human Set – Volume 1)

DUJON Bernard, PELLETIER Georges
Trajectories of Genetics

2019

BRAND Gérard
Discovering Odors

BUIS Roger
Biology and Mathematics: History and Challenges

2016

CLERC Maureen, BOUGRAIN Laurent, LOTTE Fabien
Brain-Computer Interfaces 1: Foundations and Methods
Brain-Computer Interfaces 2: Technology and Applications

FURGER Christophe
Live Cell Assays: From Research to Health and Regulatory Applications

2015

CLARYSSE Patrick, FRIBOULET Denis
Multi-modality Cardiac Imaging: Processing and Analysis

2014

CHEZE Laurence
Kinematic Analysis of Human Movement

CLARYSSE Patrick, FRIBOULET Denis
Multi-modality Cardiac Imaging: Processing and Analysis

DAO Tien Tuan, HO BA THO Marie-Christine
Biomechanics of the Musculoskeletal: Modeling of Data Uncertainty and Knowledge

FANET Hervé
Medical Imaging Based on Magnetic Fields and Ultrasounds

FARINAS DEL CERRO Luis, INOUE Katsumi
Logical Modeling of Biological Systems

MIGONNEY Véronique
Biomaterials

TEBBANI Sihem, LOPES Filipa, FILALI Rayen, DUMUR Didier, PAREAU Dominique
CO_2 Biofixation by Microalgae: Modeling, Estimation and Control

Printed and bound by CPI Group (UK) Ltd, Croydon, CR0 4YY
16/09/2026
14963082-0001